T0176158

EVOLUTIONARY BIOLOGY, CELL–CELL COMMUNICATION, AND COMPLEX DISEASE

EVOLUTIONARY BIOLOGY, CELL–CELL COMMUNICATION, AND COMPLEX DISEASE

John S. Torday, MSc, PhD
Professor of Pediatrics and
Obstetrics and Gynecology
Harbor–UCLA Medical Center
David Geffen School of Medicine
University of California—Los Angeles

Virender K. Rehan, MD
Professor of Pediatrics
Chief, Division of Neonatology
Harbor–UCLA Medical Center
David Geffen School of Medicine
University of California—Los Angeles

WILEY-BLACKWELL

A JOHN WILEY & SONS, INC., PUBLICATION

Copyright © 2012 by Wiley Blackwell. All rights reserved.

Published by John Wiley & Sons, Inc., Hoboken, New Jersey
Published simultaneously in Canada

Wiley-Blackwell is an imprint of John Wiley & Sons, formed by the merger of Wiley's global Scientific, Technical, and Medical business with Blackwell Publishing.

No part of this publication may be reproduced, stored in a retrieval system, or transmitted in any form or by any means, electronic, mechanical, photocopying, recording, scanning, or otherwise, except as permitted under Section 107 or 108 of the 1976 United States Copyright Act, without either the prior written permission of the Publisher, or authorization through payment of the appropriate per-copy fee to the Copyright Clearance Center, Inc., 222 Rosewood Drive, Danvers, MA 01923, 978-750-8400, fax 978-750-4470, or on the web at www.copyright.com. Requests to the Publisher for permission should be addressed to the Permissions Department, John Wiley & Sons, Inc., 111 River Street, Hoboken, NJ 07030, 201-748-6011, fax 201-748-6008, or online at http://www.wiley.com/go/permissions.

Limit of Liability/Disclaimer of Warranty: While the publisher and author have used their best efforts in preparing this book, they make no representations or warranties with respect to the accuracy or completeness of the contents of this book and specifically disclaim any implied warranties of merchantability or fitness for a particular purpose. No warranty may be created or extended by sales representatives or written sales materials. The advice and strategies contained herein may not be suitable for your situation. You should consult with a professional where appropriate. Neither the publisher nor author shall be liable for any loss of profit or any other commercial damages, including but not limited to special, incidental, consequential, or other damages.

For general information on our other products and services or for technical support, please contact our Customer Care Department within the United States at 877-762-2974, outside the United States at 317-572-3993 or fax 317-572-4002.

Wiley also publishes its books in a variety of electronic formats. Some content that appears in print may not be available in electronic formats. For more information about Wiley products, visit our web site at www. wiley.com.

Library of Congress Cataloging-in-Publication Data:

Torday, John S.
  Evolutionary biology, cell-cell communication, and complex disease / John S. Torday, Virender K. Rehan.
       p. ; cm.
  Includes bibliographical references and index.
  ISBN 978-0-470-64720-2 (cloth)
  1. Molecular genetics. 2. Molecular evolution. 3. Cell interaction. 4. Pathology, Molecular.
I. Rehan, Virender K. II. Title.
  [DNLM: 1. Evolution, Molecular. 2. Cell Communication. 3. Pathologic Processes. QU 475]
  QH442.T648 2012
  572.8–dc23

                                                                                    2011017542

Printed in the United States of America

10 9 8 7 6 5 4 3 2 1

Dr. Torday dedicates this book to his wife Barbara and his children Nicole Anne and Daniel Philip Torday, his daughter-in-law Dr. Erin Torday, his granddaughter Abigail Torday, and his parents Steven and Maria Torday.

Dr. Rehan dedicates this book to his parents Sain Das and Nirmala Rehan, his wife Yu Hsiu and children Amit and Anika Rehan, and his brother (the late) Dr. Sudhir Rehan.

CONTENTS

PREFACE

There have been many attempts to conceptualize and explain the process of evolution, from unicellular to multicellular organisms. Yet even in the age of genomics, we still do not even understand how novelty arises from organisms that are seemingly destined to maintain their biologic identity. Therein lays the enigma of evolution, which is only made more difficult to understand by reducing life to genes and phenotypes, using metaphoric language instead of molecular mechanisms to deconvolute the process of evolution. The challenge is to formulate a strategy for transitioning from anecdotes to a central theory of biology, and from duality to unity.

The purpose of this book is to understand the why and how of evolution by focusing on the cell as the smallest unit of biologic structure and function. The consensus is that unicellular organisms developed the complete genetic toolkit for the evolution of multicellular organisms. Multicellular organisms, in turn, devolved form and function in adaptation to their environments through metabolic cooperativity. By focusing on cell–cell communication as the organizing principle and mechanism of evolution, the deep molecular homologies that have formed the basis for vertebrate evolution can be seen simultaneously in their historic and contemporary contexts. By further focusing on lipid metabolism as an integrating mechanism linking molecular oxygen to cholesterol, overarching vertebrate evolution from the cell membrane to barrier function and neocortical evolution, we can literally connect the dots from our evolutionary past to the present and future. Using this approach, development, homeostasis, regeneration, and reproduction are seen as a structural–functional continuum, providing the means by which organisms have been able to keep apace with their ever-changing environment.

We would like to thank our mentors, who have challenged us to drill down to the essence of biology. We would also like to thank the agencies that have funded our research over the years: The National Institutes of Health, The March of Dimes, The American Heart Association, The Tobacco Research and Disease Related Program of California, The Thrasher Research Foundation, and The Los Angeles Biomedical Research Institute.

JOHN S. TORDAY
VIRENDER K. REHAN

ABOUT THE AUTHORS

John S. Torday, MSc, PhD, was born in Budapest, Hungary. He is a graduate of Boston University, with a major in biology and a minor in English. He received his master of science and doctor of philosophy degrees from the Department of Experimental Medicine, McGill University, Montreal, Quebec, Canada. He subsequently did postdoctoral training in the NIH Reproductive Biology Program at the University of Wisconsin—Madison (1974–1976). He has been a member of the faculties of Harvard Medical School (1976–1991), The University of Maryland School of Medicine (1991–1998), and The David Geffen School of Medicine (1998–present).

Virender K. Rehan, MBBS, DCH, MD, MRCP, MRCPI, was born in Shahkot, India. He obtained his initial medical training from Delhi University, India, followed by specialized training in Liverpool (England), Winnipeg (Canada), and Rhode Island (USA). He has been a member of the faculties of The Brown University School of Medicine (1995–2000) and The David Geffen School of Medicine (2000–present).

1

THE CELLULAR ORIGIN
OF VERTEBRATES

THE ORIGINS OF UNICELLULAR LIFE ON EARTH

Life has existed on Earth for billions of years, starting with primitive cells that evolved into unicellular organisms (Fig. 1.1) over the course of the first 4.5 billion years of Earth's existence. The evolution of complex biologic organisms began with the symbiotic relationship between prokaryotes and eukaryotes. This relationship gave rise to mitochondria, and the resulting diversity of unicellular

Evolutionary Biology, Cell–Cell Communication, and Complex Disease, First Edition.
John S. Torday and Virender K. Rehan.
© 2012 Wiley-Blackwell. Published 2012 by John Wiley & Sons, Inc.

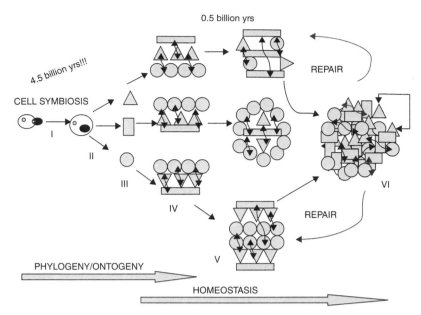

Figure 1.1. Cooperative cells as the origin of vertebrate evolution. The evolution of complex biologic oroganisms began with the symbiotic relationship between pro- and eukaryotes (**I**). This relationship gave rise to mitochondria (**II**). The resulting diversity of unicellular organisms (**III**) led to their metabolic cooperativity (**IV**), mediated by ligand–receptor interactions and cell–cell signaling. Natural selection generated increasing complexity (**V**). Failed homeostatic signaling (**VI**) recapitulates phylogeny and ontogeny, offering pathology and repair as the inverse of phylogeny and ontogeny. From Torday (2004). (See insert for color representation.)

organisms led to their metabolic cooperativity, mediated by ligand–receptor inter-actions and cell–cell signaling. Natural selection generated increasing complexity. Failed homeostatic signaling recapitulates phylogeny and ontogeny, offering pathology and repair as the inverse of phylogeny and ontogeny. How life on Earth actually began can only be speculated, unless we can witness it unfolding on "other Earths," and even then the process would probably differ from what has transpired on Earth since it is still contingent on the prevailing environmental conditions.

The question of the origins of life was first formally addressed by A. I. Oparin in 1924, and then by J. B. S. Haldane in 1929. They reasoned that if the early Earth environment lacked atmospheric oxygen, a variety of organic compounds could have been synthesized in reaction to energy from the sun, and by electrical discharges generated by lightning. Haldane suggested that in the absence of living organisms feeding on these putative organic compounds, the oceans would have attained a hot, soupy consistency.

The formation of boundaries through which things cross is the domain of cel-lular processes. Metabolic theories of the origins of life such as those of Oparin and Fox (Fox 1965) assume the existence of a primitive cell-like compartment, or protocell, in which metabolism may have emerged. Metabolism, in turn, caused the growth of the cell and its division into daughter cells when its physical limits of gas and nutrient exchange had been reached or surpassed.

One way in which cellular life has been postulated to have originated was through the well-recognized process by which the repeated wetting and drying of lipids naturally generates micelles, which are spheres composed of semipermeable membranes. Perhaps this occurred on the shores of the primordial oceans, waves depositing lipids derived from plant life at the water's edge (algae have been around for 3.5 billion years and are rich in lipids) and drying out, only to be wetted and dried again and again, repeatedly over eons. Within these primitive cells, catalytic reactions that would have reduced entropy within them could have resulted from random interactions between molecules generated by the electrical discharges during thunderstorms passing through the primordial atmosphere. In 1953, Miller and Urey tested this hypothesis experimentally by passing an electrical charge through an airtight glass reaction chamber containing water, methane, ammonia, hydrogen, and carbon monoxide, modeling the composition of the prebiotic Earth atmosphere. After days of refluxing, the apparatus was opened and the contents of the reaction vessel were analyzed. They identified a wide variety of organic compounds, including amino acids (the building blocks of proteins), sugars, purines and pyrimidines (the building blocks for DNA), fatty acids, and a variety of other organic compounds, suggesting that the conditions in the primitive Earth's atmosphere gave rise to the origins of life.

Wachtershauser (1988) refined this concept by suggesting that chemical reactions may have taken place between ions bonded to a charged surface. Advocates for this school of thought maintain that the emergence of such a structure that walled itself off from its environment by a membrane gave rise to the partitioning between life and nonlife. Membrane proponents focus on the primordial role of lipids in this process, and the fundamental role of membranes in the conversion of light energy into chemical, electrical, or osmotic energy, fostering the growth of protocells through metabolic processes within them (Morowitz 1992). Morowitz suggested that the prebiotic environment contained hydrocarbons, some of which were composed of long chains of carbon and hydrogen. These compounds accumulated on the surface of the ocean, where they interacted with minerals to generate amphiphiles, such as phospholipids, which are molecular dipoles, one end of which was hydrophilic and the other end, hydrophobic. These molecules condensed into various structures, including mono- and bilayers, or lipid sheets. Amphiphilic bilayers spontaneously form spheres in an aqueous solution, with the polar heads of the two layers pointing outward into the adjoining aqueous phase. The nonpolar ends of the bilayer point inward toward the center. This is the basic structure of biological membranes that form the outer surfaces of all cells, allowing active transport of chemicals across the membrane in conjunction with proteins interspersed in the lipid bilayer.

It has been suggested that this spontaneous formation of closed vesicles is the origin of triphasic systems consisting of a polar interior, a nonpolar membrane core, and a polar exterior, creating an interior environment. Morowitz went on to empirically demonstrate that the advent of life processes depended on these properties of amphiphilic vesicles. Nonpolar molecules such as chromophores, which absorb light energy, tend to dissolve in the nonpolar lipid core of the membrane, where the light energy is converted into electrical energy that drives various chemical reactions, including the generation of even more amphiphiles. In contemporary cells, such reactions are mediated by phosphate bond energy, whereas in their primitive condition these reactions were facilitated by pyrophosphates. The generation of new amphiphiles through this mechanism increased the vesicle size. Once the vesicle reached a critical size, it broke up into smaller, more stable vesicles in the same way that soap bubbles do. This process is thought to be the origin of cell division.

PROKARYOTES VERSUS EUKARYOTES

Eukaryotes are organisms with a membrane envelope around their nuclei like our own cells. They are assumed to have evolved from prokaryotes, such as bacteria, which lack a nucleus. Eukaryotes have numerous organelles that are absent from prokaryotes, including mitochondria, or plastids, which play an important role in energy metabolism. The unusual structure and self-replication of mitochondria and plastids had suggested to some scientists back in the nineteenth century that perhaps these structures descended from bacteria. It was subsequently determined that the mitochondrial and plastid mutations were independent of nuclear DNA. We now know that these organelles are related to bacteria, forming the basis for the *endosymbiotic theory*, which was first put forward by the Russian botanist Mereschkowski in 1905. Mereschkowski knew of the work by botanist Andreas Schimper, who had observed in 1883 that the division of chloroplasts in green plants closely resembled free-living cyanobacteria. Schimper had proposed that green plants arose from the symbiotic union of two organisms. Wallin extended this concept of endosymbiosis to mitochondria in the 1920s. At first these theories were either dismissed or ignored. More detailed electron microscopic comparisons between cyanobacteria and chloroplasts, combined with the discovery that plastids and mitochondria contain their own DNA, led to a reprise of endosymbiosis by Lynn Margulis in a 1967 paper entitled "The origin of mitosing eukaryotic cells." In her 1981 book, entitled *Symbiosis in Cell Evolution*, she postulated that eukaryotic cells originated as communities of interacting entities, including endosymbiotic spirochetes that developed into eukaryotic flagella and cilia. This last idea has not received much acceptance, because flagella lack DNA and do not show ultrastructural similarities to prokaryotes.

COEVOLUTION OF TRAITS

The most significant functional difference between pro- and eukaryotes is that prokaryotes have a rigid cell wall, while eukaryotes have a compliant plasma membrane, allowing them to easily change shape. The deformability of the plasma membrane allows for cytosis, a process by which membrane-bound vesicles inside the cell can fuse with and become part of the cell membrane. The prototype for this function is phagocytosis, or "cell eating," which allows eukaryotic cells to ingest solid particles that fuse with lysosomes containing digestive enzymes. Bacteria can feed on solid nutrients by secreting digestive enzymes into their immediate surroundings, and then absorb the molecular nutrients across the cell wall, molecule by molecule. This adaptation for feeding efficiently may have been the first step in the successful evolution of eukaryotes. This may also have been the basis for endosymbiosis by the ingestion and compartmentation of bacteria. Evidence for such a mechanism has come from studies of Archezoa, the most ancient eukaryotes. These protists have a nucleus containing chromosomes, but remain unicellular and lack mitochondria or plastids. Molecular phylogeny has documented the relationship between archezoans and eukaryotes. Another adaptation is the nuclear envelope, the structural hallmark of eukaryotes, which may have derived from invagination of the outer cell membrane. This concept is consistent with the fact that the nuclear envelope is a double membrane, and that the nuclear outer membrane is continuous with the endoplasmic membrane.

The loss of the external skeleton would have been offset by balancing selection for internal skeletal filaments and microtubules, along with mitosis, the dividing of chromosomes, and their segregation into daughter cells, since bacterial chromosomal segregation is dependent on attachment of the chromosome to the cell wall. The selection advantage is for more efficient means of feeding on solid food, and the acquisition of novel organelles. The concomitant reproductive mechanism of chromosome segregation, aided by the newly evolved microfilaments, provided a selection advantage over bacterial chromosomal replication, which must start at a single point, allowing for an exponential increase in genetic material.

(*Note*: This recurrent theme of overlapping selection for phenotypes of feeding with other complementary, coevolved adaptations is key to understanding the cellular origin of metazoan evolution. Such coevolved traits represent the emergence and contingence of the evolutionary process, all the way from molecular oxygen and cholesterol to complex physiologic traits. This concept will be repeated throughout this book.)

The loss of the prokaryotic stiff outer cell wall by eukaryotes is important in providing a possible new way of feeding, and also for other adaptations made possible by this newly evolved trait. The cytoskeleton comprises two main classes of molecules: actin filaments and microtubules. They perform complementary functions; actin filaments resist pulling forces, and microtubules resist compression and shearing forces. These properties enable the cytoskeleton to maintain the cell's form in the absence of a rigid cell wall. The cytoskeleton can also change the shape of the cell, and move things around within it. Microtubules are used for the intracellular trafficking of particles and vesicles. They also pull chromosomes apart during cell division, and are constituents of such locomotor organs as cilia and flagella. Actin filaments are active in cell division and in phagocytosis. To move things around, molecules must exert *mechanochemical activity*, meaning that they must convert chemical activity into mechanical motion. This requires that the protein molecules be able to actively change shape—in effect, they must be able to extend themselves, attach to something, and retract.

Where did this ability come from? It already existed in a rudimentary form in eubacteria. When bacteria divide, a furrow is formed in the cell membrane. This requires a mechanically active molecule; the gene sequences of certain eukaryotic mechanochemical proteins show some resemblance to this bacterial molecule. So the "fission"-aiding protein turns out to be preadapted to a cytoskeletal function.

In modern eukaryotic cells, there are several different kinds of membranes. How could they have evolved? The initial evolution of the food vacuole, budding off by endocytosis to form the cell membrane, presented just such an opportunity. In bacteria, the digestive enzymes that are secreted are synthesized by ribosomes attached to the cell membrane. If, in the original eukaryotes, a food vacuole formed randomly in response to the phagocytosis of a food particle, aided by the primitive cytoskeleton (the furrow-forming bacterial molecules), then the digestion of food within the vacuole and the ingestion of nutrients would be carried out by the original bacterial machinery.

One school of thought is that the presence of cholesterol, or related sterols, where appropriate, modified the physical properties of membranes in a manner that was crucial to the evolution of eukaryotic cells into their present form. The lipid composition of the plasma membranes of eukaryotic cells invariably includes a substantial amount of cholesterol. Konrad Bloch (1979) asked whether one could discern in the contemporary sterol pathway, and in the temporal sequence of modifying events, a directed evolutionary process operating on a small molecule and, if so, whether each step of the sequence produces a molecule functionally superior to its precursor, or molecular evolution. He demonstrated that cholesterol evolved on the appearance of oxygen in the atmosphere, facilitated by the cytochrome P450 family of enzymes necessary for cholesterol synthesis. He speculated that the biological advantage associated with cholesterol may have been due to the "reduced fluidity" or "increased microviscosity" that the addition of cholesterol imparts to the liquid crystalline state of phospholipid bilayer membranes.

Bloom and Mouritsen (Miao et al. 2002) proposed that the biosynthesis of cholesterol in an aerobic atmosphere removed a bottleneck in the evolution of eukaryotic cells. This proposed role of the physical properties of membranes in the evolution of eukaryotes is compatible with Cavalier-Smith's (2010) characterization of the evolution of eukaryotic cells, in which he identifies "twenty-two characters universally present in eukaryotes and universally absent from prokaryotes." He presents detailed arguments that, of these, the advent of exocytosis and endocytosis is the most likely to have provided the driving force for the evolution of eukaryotic cells into their present form. Bloom and Mouritsen have hypothesized that, in addition to influencing the cohesive strength of membranes, the main role of cholesterol in this evolutionary step was to relax an important constraint on membrane thickness imposed by the biological necessity of membrane fluidity—the introduction of cholesterol into phospholipid bilayer membranes increases the orientational order, but does not increase the microviscosity. Such fluidlike properties would allow large membrane curvatures without abnormal increases in permeability. As a result, with the appearance of large amounts of molecular oxygen in Earth's atmosphere between 2.3 and 1.5 billion years ago, a bottleneck in the evolution of eukaryotic cells was removed by the resultant incorporation of sterols into the plasma membrane. This interrelationship between cholesterol, membrane thickness, and cytosis is recapitulated later in this book through the overlapping of the evolution of pulmonary surfactant, the increased oxygenation of the lung, and feeding efficiency (Chapter 6). Such a reprise of a trait that served one purpose in evolution, only to serve a homologous purpose later in evolution, is referred to as an *exaptation*. Cholesterol represents a molecular phenotypic trait that has been positively selected for, beginning with unicellular organisms, all the way up through the complex physiologic properties of lung surfactant, cell–cell signaling via G-protein-coupled receptors, and endocrine regulation of physiology, all of which are catalyzed by cytochrome P450 enzymes.

CHOLESTEROL FACILITATES LIPID RAFTS FOR CELL–CELL COMMUNICATION

Plasma membrane aggregates formed by cholesterol and sphingomyelin are referred to as *lipid rafts*. Lipid rafts are composed of twice the amount of cholesterol found in the surrounding membrane, and are enriched in sphingolipids such as sphingomyelin, which is typically elevated by 50% compared to the bilayer. Phosphatidylcholine levels are decreased to offset the elevated sphingolipid levels, resulting in similar choline-containing lipid levels between the rafts and their surrounding plasma membrane. Cholesterol interacts preferentially with sphingolipids, due to their structure and the saturation of the hydrocarbon chains. Although not all phospholipids within the raft are fully saturated, the hydrophobic chains of the lipids contained in the rafts are more saturated and tightly packed than the surrounding bilayer. Cholesterol holds the raft together. Because of the rigid nature of the sterol group, cholesterol partitions preferentially into the lipid rafts where acyl chains of the lipids tend to be more rigid and in a less fluid state. One important property of membrane lipids is their amphipathic character. Amphipathic lipids have a polar, hydrophilic headgroup and a nonpolar, hydrophobic region. Cholesterol can pack between the lipids in the rafts, serving as a molecular spacer, filling gaps between associated sphingolipids.

Lipidomics is the systematic study of pathways and networks of cellular lipids. The term *lipidome* is used to describe the complete lipid profile within a cell, tissue, or organism, and is a subset of the *metabolome*, which also includes the three other major classes of biological molecules: proteins/amino acids, sugars, and nucleic acids. Lipidomics is a relatively recent research field that has been driven by rapid advances in technologies such as mass spectrometry (MS), nuclear magnetic resonance (NMR) spectroscopy, fluorescence spectroscopy, dual-polarization interferometry, and computational methods, coupled with the recognition of the role of lipids in many metabolic diseases such as obesity, atherosclerosis, stroke, hypertension, and diabetes. This rapidly expanding field is seen as a complement to the huge progress made in genomics and proteomics, all of which constitute systems biology. Yet, biology has evolved from simple lipid micelles in conjunction with that of cholesterol in response to rising oxygen tension in the atmosphere; we suggest that the lipidome may be the organizing principle for the other aspects of biology. This perspective is 180° out of phase with the conventional, descriptive way of thinking about biology and medicine, providing a novel perspective for the vertical integration of biologic data that is predictive.

Lipidomics research involves the identification and quantitation of the thousands of cellular lipid molecular species and their interactions with other lipids, proteins, and other metabolites. Investigators in lipidomics examine the structures, functions, interactions, and dynamics of cellular lipids and the changes that occur during perturbations of the system.

Han and Gross (2003) first defined the field of lipidomics through integration of the specific chemical properties inherent in lipid molecular species with a comprehensive MS approach. Although lipidomics is under the umbrella of the more general field of *metabolomics*, lipidomics is a distinct discipline because of the uniqueness and functional specificity of lipids relative to other metabolites.

In lipidomic research, a vast amount of information quantitatively describing the spatial and temporal alterations in the content and composition of different lipid molecular species is accrued after perturbation of a cell through changes in its physiological or pathological state. Information obtained from these studies facilitates mechanistic insights into changes in cellular function. Therefore, lipidomic studies play an essential role in defining the biochemical mechanisms of lipid-related disease processes by identifying alterations in cellular lipid metabolism, trafficking, and homeostasis. The growing attention on lipid research is also seen in initiatives under way such as the *lipid* *m*etabolites *a*nd *p*athways *s*trategy (LIPID MAPS consortium) and the *E*uropean *l*ipidomics *i*nitiative (ELIfe).

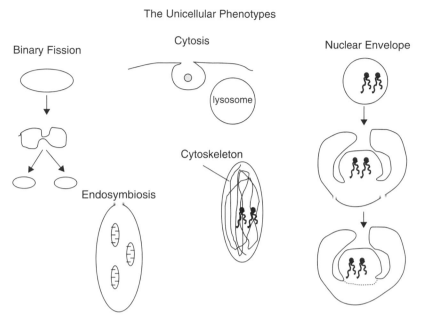

Figure 1.2. The endosymbiosis theory gives rise to specialized structures in unicellular organisms. The evolution of such structures as the mitochondria, cytoskeleton, and nuclear envelope may have resulted from cytosis.

THE ENDOMEMBRANE SYSTEM

The evolution of the thinner, more fluid eukaryotic membrane may well have accommodated the engulfment of bacteria, otherwise known as the *endosymbiosis theory* (Fig. 1.2).

The eukaryotic cell contains a variety of membranes. In bacteria, there are ribosomes attached to the outer cell membrane, synthesizing proteins that form part of that membrane—for example, proteins involved in the passage of nutrients and waste products through the membrane, and digestive enzymes secreted into the surrounding medium. It has been speculated that when a food vacuole was formed in a primitive eukaryote, it resembled the cell membrane in its ability to synthesize digestive enzymes and absorb nutrients. This is no longer true of contemporary eukaryotes—there are no ribosomes attached to the cell membrane. As a result, a food vacuole formed from the cell membrane cannot digest the food it contains. Before food can be digested, the vacuole must combine with a lysosome, which contains digestive enzymes.

If the cell membrane lacks ribosomes, and therefore the ability to synthesize proteins, how does it grow? It essentially grows by a process that is the reverse of the formation of a food vacuole—a vesicle formed within the cell fuses with the outer membrane. These vesicles originate from the endoplasmic reticulum, which is a system of double membranes within the cell. It consists of two membranes stuck to one another and formed by the flattening of spherical vesicles. Ribosomes are attached to the endoplasmic reticulum, so proteins can by synthesized. Vesicles of various kinds bud off of the endoplasmic reticulum, with the appropriate proteins embedded within them. Among these vesicles are those that fuse with the outer membrane and contribute to its growth, and also to the formation of lysosomes.

In eukaryotes there is a distinction between the cell membrane, lacking ribosomes, and the endoplasmic reticulum, which contains ribosomes and is therefore able to synthesize new cell membrane material. This may have been the first difference between these membranes as they evolved, but it has been followed by many others. The envelope surrounding the nucleus, like the endoplasmic reticulum, is formed from flattened vesicles, and so it consists of two adjacent membranes. The inner one has no ribosomes (there is no protein synthesis inside the nucleus); the outer membrane is continuous with the endoplasmic reticulum.

THE CELLULAR MECHANISM OF EVOLUTION

According to the fossil record, unicellular organisms dominated Earth for the first 4.5 billion years after this planet formed as a result of the biologic "Big Bang," adapting to their physical environment in the process. Then some 500 million years ago multicellular organisms evolved from unicellular organisms through metabolic cooperativity (see Schematic in Fig. 1.1). The cause of this cooperativity may have been due merely to the chance failure of the daughter cells to separate after mitosis (Muller and Newman 2003), or to the ongoing competition between pro- and eukaryotes, referred to as the *endosymbiosis theory* (Margulis 1981), as described above, or more likely both. The subsequent symbiotic adaptation of bacteria, leading to the formation of mitochondria, provided the energy source for the emergence of multicellular organisms. Metabolic cooperativity, and the evolution of multicellular organisms was facilitated by the production of paracrine factors for cell–cell signaling. The target cell was near the signaling cell that mediated homeostasis in primitive multicellular organisms. Such paracrine mechanisms were subsequently exploited by eukaryotes in the emergence of multicellular organisms.

At this point, we should mention that specific unicellular traits were mimicked by multicellular organisms. The progressive increase in size was a survival strategy (Bonner 2000), in tandem with host defense mechanisms such as the nuclear envelope to safeguard and protect the increasing surface area. Such adaptations were undoubtedly fostered by the functionally linked increases in oxygenation and cellular metabolism, although this would have been limited by nutritional requirements, necessitating the evolution of mitochondria. At this stage in evolution, multicellularity may have evolved to accommodate the combined selection pressures of the size–oxygenation–metabolism demands. After all, unicellular organisms had already immortalized themselves through the mechanism of binary fission. In response, multicellular organisms devised a way of immortalizing themselves through asexual and sexual reproduction.

WHY EVOLVE?

With these concepts for the beginnings of life on Earth in mind, why would the process of evolution be necessary? To understand this mechanism, please refer to Figure 1.3.

As alluded to above, the formation of lipid micelles as semipermeable membranes would have generated an internal space in which catalysis would have reduced the entropy of the molecules within it. Over eons, these primitive cell-like entities would have adapted to their physical environment through communication, during which nucleic acids would have emerged to form the database for

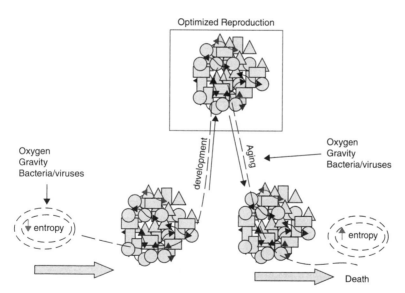

Figure 1.3. Evolutionary selection pressure, development, homeostasis, and aging. The schematic depicts the evolution of vertebrates, starting with the generation of micelles (far left), or semipermeable membrane spheres in which entropy decreased as a result of catalysis. Selection pressure due to external forces (oxygen, gravity, bacteria, viruses) gave rise to multicellular organisms through cell communication (depicted by arrow between cells). Cell–cell communication evolved into mechanisms of development, homeostasis, and repair, optimized (red box) for reproduction. Eventually this system fails as payback for defying the second law of thermodynamics, shunting energy allocation toward reproductive success, resulting in decreased cell communication (aging) and increased entropy (death). (See insert for color representation.)

storing biologic information, allowing the transfer of information from one generation to the next. This, and the advent of multicellular organisms made possible by cell–cell communication, have facilitated metabolic cooperativity. Therefore, the essence of evolution can be reduced to forms of communication: unicellular communication with the environment, cell–cell communication for multicellularity, and the communication of biologic information from generation to generation, or reproduction.

There is evidence for the continued existence of some of these properties in present-day organisms. The unicellular stage of slime molds, for example, is as free-swimming amoebas. But when there is a lack of food in the environment, the amoebas will form colonial organisms, signaling to one another through the generation of the chemical second messenger cyclic adenosine monophosphate. Similarly, sponges exist in an independent, unicellular form, generating the familiar bath sponge colonial organisms during their lifecycle. Importantly, Nicole King has shown that the unicellular form of the sponge has all of the genes necessary to form the colonial sponge, providing evidence that single-celled organisms probably evolved the metazoan toolkit (King et al. 2003).

Yet, this still does not explain the mechanism(s) of evolution. To answer that question, we can go back to the wetting and drying of lipids to form micelles, the wetting due to wave action, and the drying due to the heat from the sun. The drying would have been further affected by the 23.45° degree tilt of Earth's axis, causing seasonal variations, or the environmental "bias" (skewness) that resonates throughout animal life, as the determinant of evolutionary change according to Wallace Arthur (2004). *It is that resonance that this book seeks to identify.*

Moreover, the dynamic nature of Earth's atmosphere is due to the interaction between its organic and inorganic components. This is why evolution was necessary for life on Earth—consider a thought experiment in which an organism was perfectly adapted to its environment—that organism would have become extinct over generations because the environmental conditions that initiated and perpetuated life on Earth are constantly changing, requiring a mechanism for life to be able to adapt to such changes. Perhaps this is the mechanistic principle behind Darwin's metaphoric survival of the fittest.

CELL–CELL COMMUNICATION AND AGING

Organisms have survived because they have devised adaptive genomes that allow them to change in response to the ever-changing nature of Earth's environments (please refer to Fig. 1.3). This has come in the form of their reproductive strategy, which is optimized to generate the largest number of offspring suited for the environment into which they are born. This comes at a cost, because the energy of reproduction is selected to optimize the organism's internal physiologic milieu, namely, the cell–cell signaling mechanisms of both somatic cells and gametes.

But that energy debt must somehow be repaid because the Second Law of Thermodynamics cannot be violated—the first and second laws of thermodynamics state that the total energy content of the universe is constant, and that total entropy is continually increasing.

This assumes that there is a finite amount of energy during the lifecycle. Hayflick (2007) has unequivocally stated that longevity is genetically determined, whereas aging is epigenetic. Therefore, by definition, there must be a finite amount of energy generated during the lifecycle of any organism, which is then distributed throughout the period between birth and death in response to selection pressure for reproductive success. As a result, the bioenergetics are optimized during the reproductive phase, followed by a progressive loss of energy during the postreproductive phase of life, leading to the breakdown in cell–cell communication, aging, and ultimately death, as a result of the progressive increase in entropy.

This mechanistic explanation for the process of aging is consistent with descriptive theories of aging such as the mutation theory, antagonistic pleiotropy, and the disposable soma, as follows.

The Mutation Accumulation Theory of Aging

Peter Medawar (1952) thought that aging resulted from neglect. Since nature is highly competitive, and virtually all animals in nature die before they age, there is no selection advantage for traits that maintain life beyond the phase of the lifecycle when most animals would be dead, anyway, killed by predators, disease, or accidentally.

This theory, referred to as *mutation accumulation,* proposes that there are random, maladaptive mutations that occur late in life, unlike most deleterious mutations that are lost through the process of natural selection. As a result, they accumulate over time, causing the deterioration and damage that are conventionally attributed to aging. The mutation accumulation theory suggests that evolution is a function of the age at which the organism is able to reproduce. Traits adversely affecting an organism before that age would severely compromise the organism's ability to pass such characteristics on to its offspring. Therefore, such traits would be selected against. Characteristics that cause the same adverse effects beyond the reproductive phase would have relatively little effect on the organism's ability to reproduce, and therefore might be tolerated by natural selection. This concept fits well with the observed diversity in mammalian lifespans (and differing ages of sexual maturity), and is relevant to all of the other theories of aging discussed below.

The Antagonistic Pleiotropy Theory of Aging

The mutation accumulation theory of aging was extended by George Williams (1957), who proposed that senescence might directly cause death. Senescence would compromise the animal's ability to fight off predators, rendering them susceptible, whereas the younger animals would have been able to escape. Or senescence could undermine the animal's immunity, rendering it vulnerable to fatal infection. Williams thought that because nature was so highly competitive, due to the evolutionary process, even the smallest senescent changes in fitness would put one at risk in the wild.

Williams proposed the theory of antagonistic pleiotropy. *Pleiotropy* is defined as one gene having more than one phenotypic effect. *Antagonistic pleiotropy* means that some effects of the gene are beneficial while others are not. Such genes are selected for their beneficial effects early in life, but they then become deleterious in later life. Since evolution selects for reproductive success, the increased fertility could be selected for, but could result in premature death. Therefore aging is a natural consequence of adaptive physiology.

Antagonistic pleiotropy continues to be a popular and testable theory of aging. Williams' concept of aging has been validated by more recent studies of animal populations, in which senescence contributes substantially to the death rate. These studies, along with genomic studies, have undermined the mutation accumulation theory since the genes that cause aging are not random mutations. The aging genes are highly conserved across yeast, worms, fruit flies, and mice.

An inherent problem with antagonistic pleiotropy and other theories that assume that aging is an adverse side effect of some beneficial function is that the linkage between adverse and beneficial effects would need to be *rigid* in the sense that the evolutionary process would not be able to evolve a way to accomplish the benefit without incurring the adverse effect, even over a very long timespan. Such a rigid relationship has not been demonstrated experimentally and, in general, evolution is obviously able to independently and individually adjust myriad organismal characteristics.

The Disposable Soma theory of Aging

First proposed by Kirkwood (1977), the *disposable soma theory* states that the body budgets the amount of energy available to it. The body uses nutrient energy for metabolism, reproduction, repair, and maintenance. Because of the limited supply of nutrients, the body compensates by doing none of these things optimally. By limiting the amount of energy used for repair, the body gradually deteriorates with age. The disposable soma theory is attractive because it seems intuitively correct, yet there is evidence that refutes this idea. Whereas lack of food should hasten aging, food restriction experiments have shown that experimental animals live longer. It is true that restricting food intake would slow metabolism, decreasing the damage done by free radicals, while the repair process remains unaffected. But dietary restriction has not been shown to increase lifetime reproductive success (fitness), because when food availability is low, reproductive output is also low. So food restriction does not completely disprove the disposable soma theory.

Experimentally, some animals lose fertility when their lifespan is protracted by food restriction, while others are unaffected. Males typically remain fertile when they are food-restricted, whereas females do not. Moreover, the females present a paradox because their loss of fertility does not correlate well with the increase in lifespan.

The Cell–Cell Signaling Model of the Lifecycle and Aging

The success of the cell–cell interaction mechanism is in its proven ability to communicate knowledge acquired during the lifecycle from one generation to the next in adaptation to the ever-changing environment. As shown in Figure 1.3, agents that initially fostered evolution, such as oxygen, gravity, and bacteria, are omnipresent in the environment, but through the evolving reproductive strategy, organisms have adapted to their ever-changing environment. However, those same agents that gave rise to the evolutionary strategy continue to affect the organism throughout its lifecycle, so the system must fail, that is, age and die, as a result of the breakdown in cell–cell signaling as the mirror image of development and homeostasis. This phenomenon has been described as *molecular fidelity and antagonistic pleiotropy*. The difference between these metaphoric descriptions of aging and the model that we are proposing in Figure 1.3 is that the mechanistic hypothesis that the ligand–receptor mechanisms of development fail during biologic aging is testable and refutable. The implications of such a mechanism are manifold, not the least of which is the confusion between aging and cumulative pathology, as has been pointed out by Michael Rose (2009). Moreover, by determining how evolution has been optimized for reproductive success at the cell–molecule level, we can apply those same mechanisms to the postreproductive period to actively and effectively promote healthy aging based on tried and true evolutionarily adaptive biological mechanisms, rather than on *ex post facto* reasoning and results-oriented pharmacology.

As mentioned earlier, studies by King et al. (2003) have indicated that unicellular organisms had the entire genetic complement to form multicellular organisms, suggesting that the principles of evolution go all the way back to one-celled organisms. In the following chapters we will show how unicellular organisms generated multicellular organisms through cell–cell molecular coupling mechanisms acting as evolutionary transducers. As a result, the mechanisms of evolution are continuous from uni- to multicellular organisms; we will refer to those highly conserved mechanisms as the evolutionary *data operating system*.

In Chapter 2 we will describe the cellular–molecular basis for lung alveolar development and homeostasis, which forms the basis for our evolutionary approach to the physiology of the lung as the platform for other evolved physiologic traits. During the process of determining the cellular–molecular basis for alveolar physiology, we have come to the realization that cell–cell signaling mechanisms are far more likely to have evolved under Darwinian selection pressure than as a result of random genetic mutations.

REFERENCES

Arthur W (2004), *Biased Embryos and Evolution*, Cambridge University Press, Cambridge, UK.

Bloch KE (1979), Speculations on the evolution of sterol structure and function. *CRC Crit. Rev. Biochem.* 7(1):1–5.

Bonner JT (2000), *First Signals: The Evolution of Multicellular Development*, Princeton University Press, Princeton, NJ.

Cavalier-Smith T (2010), Origin of the cell nucleus, mitosis and sex: roles of intracellular coevolution. *Biol. Direct.* 5:7.

Fox SW (ed.) (1965), *The Origins of Prebiological Systems and of Their Molecular Matrices*, Academic Press, New York.

Han X, Gross RW (2003), Global analyses of cellular lipidomes directly from crude extracts of biological samples by ESI mass spectrometry: a bridge to lipidomics. *J. Lipid Res.* 44(6):1071–1079.

Hayflick L (2007), Entropy explains aging, genetic determinism explains longevity, and undefined terminology explains misunderstanding both. *PLoS Genet.* 3(12):e220.

King N, Hittinger CT, Carroll SB (2003), Evolution of key cell signaling and adhesion protein families predates animal origins. *Science* 301(5631):361–363.

Kirkwood TB (1977), Evolution of ageing. *Nature* 270(5635):301–304.

Margulis L (1981), *Symbiosis in Cell Evolution.* W. H. Freeman, New York.

Medawar PB (1952), *An Unsolved Problem of Biology.* H.K. Lewis & Co., London.

Miao L, Nielsen M, Thewalt J, Ipsen JH, Bloom M, Zuckermann MJ, Mouritsen OG (2002), From lanosterol to cholesterol: structural evolution and differential effects on lipid bilayers. *Biophys. J.* 82(3):1429–1444.

Morowitz HJ (1992), *Beginnings of Cellular Life: Metabolism Recapitulates Biogenesis.* Yale University Press, New Haven, CT.

Muller GB, Newman SA (2003), *Origination of Organismal Form: Beyond the Gene in Developmental and Evolutionary Biology.* The MIT Press, Cambridge, MA.

Rose MR (2009), Adaptation, aging, and genomic information. *Aging* 1(5):444–450.

Torday JS (2004), A periodic table for biology. *Scientist* 18(12):32–33.

Wachtershauser G (1988), Before enzymes and templates: theory of surface metabolism. *Microbiol. Rev.* 52:452–484.

Williams GC (1957), Pleiotropy, natural selection and the evolution of senescence. *Evolution* 11:398–411.

2

REDUCING LUNG PHYSIOLOGY TO ITS MOLECULAR PHENOTYPES

HORMONAL ACCELERATION OF LUNG DEVELOPMENT

The impetus for our interest in the cellular–molecular basis for the evolution of the lung derives from reductionist studies of lung development in our laboratory and that of many others over the last >50 years.

Prior to that time, it was known that lung development was the result of epithelial–mesenchymal interactions, mediated by small molecules that passed between these two embryonic layers to generate the 40 cell types of the fully differentiated lung. Grobstein (1967) had shown experimentally that the lung could continue to develop spontaneously in chemically defined tissue culture medium. More importantly, if he separated the two embryonic tissue layers, the endoderm and mesoderm, the isolated tissues failed to develop. However, if he placed a

Evolutionary Biology, Cell–Cell Communication, and Complex Disease, First Edition.
John S. Torday and Virender K. Rehan.
© 2012 Wiley-Blackwell. Published 2012 by John Wiley & Sons, Inc.

semipermeable membrane between the two tissue layers, the lung would continue to develop, indicating that small molecules passing between the two layers mediated this process. The nature of those molecules remained a mystery until a breakthrough in this field occurred in (1969), when Liggins discovered that the adrenocortical hormone cortisol stimulated structural and functional fetal lung development, indicating that the process was not genetically determined, but that there were epigenetic mechanisms that controlled this process.

While studying the effects of cortisol on the birth process in sheep, Liggins noticed that the newborn preterm lambs that were hormonally induced to give birth were able to breathe on their own, which was astonishing given that these animals would have died if they had been delivered by Cesarean section. This hormonal effect on the process of lung development was subsequently reproduced in a number of laboratories around the world, including that of Mary Ellen Avery's research group at McGill University, Montreal, Quebec, Canada, with whom one of the authors of this book, John Torday, collaborated while he was a graduate student at McGill. He subsequently began studying the cellular mechanism for cortisol's effects on lung development, particularly the production of lung surfactant, which is the key physiologic process for the adaptation of the lung to air breathing. Barry Smith and John Torday initially showed that cortisol directly stimulated the synthesis of surfactant in cultured fetal rabbit and human lung cells. Smith subsequently made the seminal observation that the hormone did not directly stimulate surfactant synthesis by the epithelial type II cells that synthesize the surfactant, which would have been logical, but instead stimulated the production of a low-molecular-weight protein by the neighboring fibroblasts. This was the first molecular evidence for the existence of a specific agent mediating cell–cell communication in normal lung development. The other functional property of developing lung fibroblasts known at that time was that those fibroblasts directly adjacent to the epithelial type II, referred to as *lipid-laden fibroblasts*, or *lipofibroblasts*, stored neutral lipids, although their functional significance was unknown. Twenty years later, the Torday laboratory discovered that those stored lipids served two important physiologic functions that would elucidate well-known but poorly understood phenomena regarding lung physiology and disease:

1. The lipids protected the alveoli against oxidant injury, consistent with observations by Frank and Roberts that newborn mice, rats, and rabbits, in which lipofibroblasts are prevalent, were much more resistant to oxygen injury than their adult counterparts.

2. The neutral lipids stored in the lipofibroblasts functioned as substrate for surfactant phospholipid synthesis by the neighboring epithelial cells that were actively transported, explaining how the lung was able to rapidly generate more surfactant on demand when the alveoli were stretched during the inspiration of air. Faridy (1976) was the first to report this effect of alveolar distension on surfactant production in ventilated dogs. This was followed

by Nicholas and Barr's observation that when rats expanded their lungs by forcing them to swim, they rapidly produced more surfactant. Other studies of fetal lung development showed similar effects of alveolar fluid distension in the womb on surfactant production.

NEUTRAL LIPID TRAFFICKING AND LUNG EVOLUTION

The mechanisms involved in the stretch-activated increase in lung surfactant production were not determined until 1995 (see Fig. 2.1), when the Torday laboratory discovered the principle of *neutral lipid trafficking*, which mediates the transfer of lipid from the alveolar circulation to the interstitial lipofibroblast, and from the lipofibroblast to the alveolar type II cell for facilitated surfactant production. We found that cultured fetal lung fibroblasts actively accumulate serum neutral lipid from the surrounding medium, but interestingly, that they did not release it; in contrast to this, cultured alveolar epithelial type II cells were unable to take the neutral lipid up. However, when the lipid-filled fetal lung fibroblasts were cocultured with the lung epithelial cells, the neutral lipid was rapidly and specifically

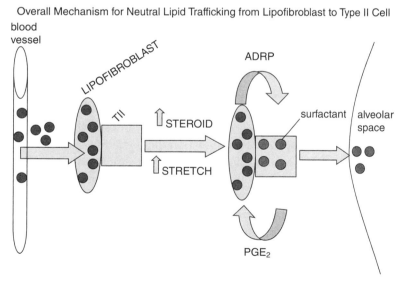

Figure 2.1. Neutral lipid trafficking from lipofibroblast to type II cell. Lipofibroblasts actively take up and store neutral lipid from the circulation by expressing adipocyte differentiation–related protein (ADRP). Alveolar type II cells (TII) recruit these neutral lipids by secreting prostaglandin E_2. Hormones and mechanical stretching coordinately regulate these mechanisms to integrate distension of the alveoli during breathing with the production of lung surfactant. (From Torday et al. 1995.) (See insert for color representation.)

incorporated into surfactant phospholipids by the epithelial type II cells. The transfer of the neutral lipid was found to be cortisol-stimulated, substantiating the theory that this was a regulated mechanism. The paradoxical active transfer of lipid from lipofibroblasts to alveolar epithelial type II cells, but the inability of the lipofibroblasts to release the lipids in cell culture was resolved when it was discovered that exposing the fibroblasts to culture medium containing secretions from the epithelial type II cells caused release of the neutral lipid, indicating that some soluble factor(s) secreted by the epithelial cells caused the release of the neutral lipid from the lipofibroblasts. Subsequent chemical analysis of the secretions of the alveolar epithelial type II cells revealed that they secreted a lipid-soluble substance that caused the release of triglyceride from the fibroblasts. On further examination, it was found that these cells produce prostaglandins E_2, a highly biologically active lipid mediator of secretion, and that there were specific prostaglandin E_2 receptors on the lipofibroblasts. Experimentation revealed that the prostaglandin E_2 produced by the alveolar epithelial type II cells caused the active release of the neutral lipid from the lipofibroblasts. This signaling from the epithelium to the fibroblast for lipid substrate mediated by prostaglandin E_2 acquired greater developmental physiologic significance when it was found that the production of prostaglandin E_2 by alveolar epithelial type II cells increased severalfold as birth approached, peaking just before delivery, and that both cortisol and the stretching of the type II epithelial cells increased prostaglandin E_2 synthesis and secretion. Taken together, these observations suggested cellular cooperativity by which neutral lipids were actively recruited from the lung alveolar circulation and stored by the lipofibroblasts, and subsequently actively "trafficked" to the alveolar epithelial type II cells for surfactant phospholipid synthesis in preparation for air breathing. These findings, combined with the earlier observation that cortisol stimulated the overall transfer of lipid from the lipofibroblast to the alveolar epithelial type II cell, provided evidence of such a regulated, cell–cell interactive process.

We subsequently discovered (Fig. 2.2) that the developing lipofibroblasts obtain the neutral lipid from the circulation by producing adipocyte differentiation–related protein (ADRP), a molecule that is necessary for the uptake and storage of lipid by all cells. ADRP encapsulates the lipid droplets for storage and secretion, and is required for the subsequent uptake and incorporation of the lipid into phospholipids by the alveolar epithelial type II cells.

The neutral lipid trafficking mechanism for regulation of alveolar surfactant provided phenotypic functional molecular markers for our subsequent study of the developmental mechanisms that determine stretch-induced, on-demand surfactant production. We discovered that parathyroid hormone–related protein (PTHrP) (refer to Fig. 2.1) is a stretch-regulated protein produced by the lung alveolar epithelial type II cells, which stimulates surfactant synthesis through a cell–cell interactive mechanism: PTHrP binds to its cell surface receptor on the lung lipofibroblast, increasing both (1) lipid uptake by stimulating ADRP and (2) leptin, the secretory paracrine product of the fibroblast that stimulates alveolar

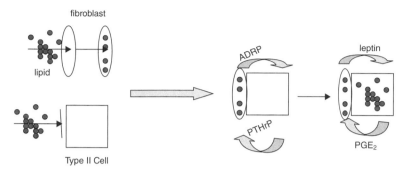

Figure 2.2. Experimental evidence for neutral lipid trafficking. Monolayer cultures of lung fibroblasts actively take up neutral lipid, but don't release them (top left) unless they are cocultured with type II cells. Type II cells cannot take up neutral lipid (lower left) unless they are cocultured with lung fibroblasts. Fibroblast uptake of neutral lipid is determined by PTHrP from the type II cell, which stimulates ADRP expression by the lung fibroblast (middle); stretching cocultured lung lipofibroblasts and type II cells increases surfactant synthesis by coordinately stimulating PGE_2 production by type II cells, causing release of neutral lipid by the lipofibrfoblasts, and leptin secretion by the lipofibroblasts, which stimulates surfactant phospholipid synthesis by the type II cells. (See insert for color representation.)

epithelial type II cell surfactant synthesis. Expression of both leptin and the PTHrP receptor by the lung fibroblast, and PTHrP, prostaglandin E_2 (PGE_2), and the leptin receptor by the alveolar epithelial type II cell are all stretch-regulated mechanisms. This results in the neighboring fibroblasts and type II cells coordinately mediating the increase in lung surfactant production through integrated, stretch-regulated increases in PTHrP and leptin specifically signaling through their mutual cell surface receptors. This mechanism is further facilitated by the stretch-stimulated synthesis and release of PGE_2 by the alveolar epithelial type II cells, causing release of the neutral lipid from the lipofibroblasts, ensuring the availability of lipid substrate for stretch-mediated surfactant phospholipid synthesis. Moreover, PTHrP is a potent blood vessel dilator that stimulates alveolar capillary perfusion. Taken together, the coordinate stretch-mediated effects of PTHrP, leptin, and PGE_2 and their complementary target cell receptors account for the mechanism of ventilation/perfusion matching, which is the physiologic principle for alveolar homeostasis.

It is hard to imagine how such an exquisitely integrated mechanism for the coordinate physiologic regulation of surfactant phospholipid could have occurred merely by chance, given the variety of cell types and genetically determined ligand–receptor-mediated signaling mechanisms involved, their close proximity to one another, and their molecular regulation by endocrine hormones and paracrine factors, all regulated by the effects of stretch in support of alveolar homeostasis for gas exchange. If you were to calculate the probability that these events occurred by chance, you would multiply the length of time it took to form the specialized lipofibroblasts of the mammalian lung by the length of time it took to evolve the coelom-like epithelium-lined alveoli. Even by the crudest of estimates, mammals took more than 4 billion years to evolve, and coelomic cavities lined with epithelia took more than 4 billion years to evolve. Multiplying 4×10^9 by 4×10^9 equals 16×10^{18} years, which is older not only than the age of Earth itself but also than the estimated age of the universe!

Alternatively, it is more than reasonable to hypothesize that this mechanism was the result of positive Darwinian selection pressure for surfactant production mediated by cell–cell interactions. This physiologic interrelationship is key to our understanding of the evolutionary positive selection pressure for this process—in order for the alveoli to have developed a progressively smaller diameter, increasing the surface area/blood volume ratio for gas exchange, surfactant production also had to become progressively more highly regulated, based on the the law of Laplace, which states that the surface tension of a sphere (i.e., alveolus) is inversely related to its diameter. It also formed the basis for the structural evolution of the alveolar wall through the modifications in the mesenchymal fibroblast and epithelial cell populations, as discussed above. This evolutionary concept is supported by the consistent observation that surfactant produced by the epithelial lining cells of the gas exchangers of fish, amphibians, reptiles, mammals, and birds increases progressively in response to hormonal and stretch regulation over both phylogenetic and developmental time in close association with the progressive decrease in alveolar diameter.

Mechanistically, the transition from the regulation of surfactant by so-called housekeeping genes to the cell–cell interactive mechanisms just described required the sequential evolution of *cis* regulatory control of surfactant synthesis—the PTHrP, leptin, and prostaglandin E_2 signaling pathways all act via second messengers such as cyclic AMP and phosphatidylinositol, which interact with nuclear transcription factors to regulate surfactant synthesis and secretion. Transition from housekeeping to *cis* regulatory mechanisms in adapting to the environment are a recurrent theme in evolutionary biology, although they haven't been looked at longitudinally in the way that we have systematically elucidated lung evolution.

This cell–cell molecular model of lung evolution forms the basis for the subsequent chapters of this book that delve into the experimental evidence for the evolutionary strategy represented by this cell communication mechanism, particularly how the cellular communication mechanisms facilitated the ontogenetic and phylogenetic increases in the alveolar surface area of the lung to accommodate the metabolic drive caused by the atmospheric increases and decreases in oxygen tension. Once the cell–cell communication platform for lung evolution is established, we extend the model to whole-animal physiology, based largely on the evolution of the adaptive fat cells of the lung (the lipofibroblasts providing substrate for surfactant production), and the evolution of the systemic fat cells as the basis for endothermy, as mechanistic keys to understanding the evolution of land vertebrates.

OTHER EXAMPLES OF CELLULAR COOPERATIVITY

The Mammary Gland

The lung and mammary gland share developmental and homeostatic mechanisms that may help in understanding the evolution of each of these structures. Like the lung, the mammary gland also develops as a result of mesenchymal–epithelial interactions—and also like the lung, deletion of the PTHrP gene inhibits mammary formation. As mentioned above, ADRP is necessary for the trafficking of neutral lipid from the circulation to the lipofibroblast, and from the lipofibroblast to the alveolar type II cell for surfactant synthesis. Similarly, ADRP and other perilipin/ADRP/TIP47 (PAT) family proteins have been implicated in the formation of milk in the mammary gland. ADRP-coated lipid droplets are stored in adipocyte-like structural cells adjacent to the mammary epithelial cells that line the milk ducts, also called *alveoli*. ADRP appears to mediate the transit of milk from the structural cells to the epithelium, in much the same way as the lipid substrate for surfactant is trafficked to the alveolar epithelial cell.

Hepatocyte–Hepatic Stellate Cell Cooperativity

Like our delineation of the metabolic interactions between the alveolar type II cell and the lipofibroblast, the liver hepatic stellate cell and hepatocyte show similar metabolic cooperativity. These observations were facilitated by the introduction of effective methods for the isolation of parenchymal liver cells, namely, hepatocytes, and nonparenchymal hepatic stellate cells. Coculture of these two cell types has provided evidence for the existence of intercellular communication mediated mainly in a paracrine fashion by the release of various cellular mediators, giving rise to the liver lobule.

The communication between cells of the liver lobule takes place via three basic mechanisms: gap junctions, paracrine, and juxtacrine (which requires close apposition of cell membranes). A classic example of cell–cell communication in support of metabolic cooperativity in the liver is the metabolism and storage of retinoids. Retinoids must be provided in the diet, either as vitamin A or as provitamin A carotenoids, since they are necessary for embryonic development, growth, vision, and survival of vertebrates.

Vitamin A is absorbed in the small intestine, where it is incorporated into chylomicrons as retinyl esters for release into the lymph, and further distributed via the circulation to the liver for storage. Chylomicron remnants containing retinyl esters are internalized exclusively by hepatocytes. Very shortly after receptor-mediated endocytosis, retinyl esters present in chylomicron remnants are hydrolyzed to retinol in early endosomes. Retinol combines in the endoplasmic reticulum with retinol-binding protein, and is then either secreted from the hepatocyte, oxidized to retinoic acid, or stored as retinyl esters. Retinol binds intracellularly to specific cellular retinal-binding proteins, CRBP 1 and II, which also play an important role in retinyl ester hydrolysis and formation, and oxidation of retinal to retinoic acid. Most of the retinol in rat liver is stored in hepatic stellate cells (70–95%).

In the rat, the amount of retinoid present in stellate cells and hepatocytes is dependent on cell–cell interactions between these cell types. These interactions are regulated by a variety of molecules, ranging from prostanoids to growth factors and growth factor receptors. Many systemic mediators affect the liver during inflammation, resulting in increased glucose output by the liver, yet these same mediators do not stimulate glucose secretion by isolated hepatocytes. The release of glucose in response to inflammatory mediators is blocked by cyclooxygenase inhibitors, suggesting that the effect is due to indirect stimulation by prostanoids synthesized in nonparenchymal liver cells. Prostanoids have been shown to stimulate glycogenolysis by isolated hepatocytes. It has more recently been shown that like the lung and mammary gland lipofibroblasts, the hepatic stellate cells also express ADRP, which facilitates retinol uptake by these cells and stabilizes the phenotype.

SUMMARY

In this chapter we have formed a conceptual mechanistic link from the evolution of unicellular organisms to multicellular organisms through cell–cell communication described in Chapter 1, to the cell–cell communication mechanisms that form the basis for the mammalian lung. In Chapter 3, we will use the cell-molecular coupling mechanisms for alveolar homeostasis presented in the current chapter as the basis for an integrated approach to understanding the evolution of the lung.

REFERENCE

Faridy EE (1976), Effect of distension on release of surfactant in excised dogs' lungs. *Respir. Physiol.* 27(1):99–114.

Grobstein C (1967), Mechanisms of organogenetic tissue interaction. *Natl. Cancer Inst. Monogr.* 26:279–299.

Liggins GC (1969), Premature delivery of foetal lambs infused with glucocorticoids. *J Endocrinol.* 1969 Dec;45(4):515–23.

Nicholas TE, Power JH, Barr HA (1982), Surfactant homeostasis in the rat lung during swimming exercise. *J. Appl. Physiol.* 3(6):1521–1528.

Torday JS, Hua J, Slavin R (1995), Metabolism and fate of fetal rat lung fibroblast neutral lipids. *Biochim. Biophys. Acta* 1254:198–206.

3

A CELL–MOLECULAR STRATEGY TO SOLVING THE EVOLUTIONARY PUZZLE

RATIONALE FOR CELL–MOLECULAR EVOLUTION

The greatest challenge for biology is to disentangle the mechanisms of evolution, and determine its core principles. This effort began in earnest with Darwin's publication of *The Origin of Species* in 1859. Darwin was not the first or the only one to suggest that biology has evolved. But what set him apart from the rest was that

Evolutionary Biology, Cell–Cell Communication, and Complex Disease, First Edition.
John S. Torday and Virender K. Rehan.
© 2012 Wiley-Blackwell. Published 2012 by John Wiley & Sons, Inc.

he was a master at exhaustively describing the varieties of plants and animals that he observed while serving as a naturalist during a 5-year voyage on *HMS Beagle* (1831–1836). His descriptions were epic in proportion, comparable to those of the Bible, which may have been his intent, whether consciously or unconsciously. He produced a perspective on the origins of *Homo sapiens* that challenged "the great chain of being," which was the prevailing view humankind's place in the universe at that time. The chain started with God, progressing downward to angels, demons, the stars, the moon, kings, princes, nobles, civilians, wild animals, domesticated animals, trees, other plants, gems, precious metals, and other minerals. Darwin's genius was in his exploitation of the common practice of domestication and breeding of dogs, pigeons, and livestock as working examples of the underlying principles of evolutionary biology. By using such everyday examples that people were familiar with, he won the British public over, and he has largely been successful throughout the world. The one glaring exception is the United States, where some people still believe in creationism.

Darwinian evolution is perhaps the most important conceptual breakthrough in the history of human thought. In the interim, since Darwin's publication of *The Origin of Species* in 1859, we have gained knowledge of developmental biology and Mendelian genetics. The merging of these disciplines with Darwinian evolution has formed the *modern synthesis*, despite which we still do not understand how genes produce evolutionary novelty.

One of the biggest obstacles to this effort has been the assumption that the only scientific basis for the evolutionary process is in the hard evidence provided by the fossil record. This is particularly problematic when looking for the molecular regulatory changes that brought about developmental and physiologic novelties, which are difficult enough to find using molecular biologic techniques, let alone fossil bones. Alternatively, we have postulated that the residual evidence for the processes of vertebrate lung evolution, for example, is embedded in contemporary developmental, physiologic, and regenerative gene regulatory networks (GRNs), as displayed in Figure 3.1, which determine lung form and function, particularly when they are viewed from a comparative, developmental perspective. These are what Konrad Bloch (1979), who discovered how cholesterol is synthesized, termed *molecular fossils*. Briefly (refer to Fig. 3.1), the metabolic drive for oxygen put pressure for selection on the evolution of lung structure and function, resulting in the progressive thinning of the alveolar wall, decreasing alveolar size, and increasing production of lung surfactant. Each of these processes is represented by one of the lines in the family of parallel lines depicted in Figure 3.1 that together generated the evolution of the mammalian lung. In the absence of the complete gene regulatory network for any one of the processes represented by these parallel lines, segments of each line, represented by the broken lines, could be reassembled to approximate the processes of lung evolution.

This same approach could be used to determine the evolutionary mechanisms underlying other physiologic processes as well, given adequate developmental,

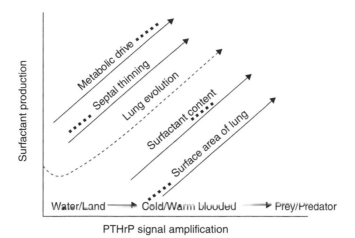

Figure 3.1. Major trends in lung evolution plotted against epochs in vertebrate evolution. Briefly, the metabolic drive for oxygen imposes pressure for selection on the evolution of lung structure and function, resulting in thinning of the alveolar septa, decreasing alveolar size, and increasing the production of surfactant. Each of these processes is represented by one of the family of parallel lines that together generated the evolution of the mammalian lung. Therefore, in the absence of the complete gene regulatory network for any one of these parallel lines, segments of each, represented by the dashed lines, can be used to trace the process of lung evolution.

and comparative genomic data. Adopting this approach would be an inducement to generate such mechanistic, developmental, and comparative data in the future. In contrast to this, evolutionary biology is dominated by descriptive observations and associations, based on mechanisms of genetic mutation and natural selection, which do not inspire hypothesis testing science. By tracing the changes in these gene regulatory networks to determine a molecular phenotype developmentally across species, you can see how cell–cell signaling has generated form and function in support of the physiologic process. We have used this approach to deconstruct the evolution of the mammalian lung by reverse-engineering the cell–cell interactions that have resulted in evolution of the *cis* regulation of lung surfactant (Torday and Rehan 2004).

We have chosen to focus on lung surfactant for several reasons: (1) its surface tension-reducing activity was necessary for the sequential phylogenetic and ontogenetic decreases in alveolar surface area to accommodate increased oxygenation; (2) it is necessary for survival; (3) its development and phylogeny are well described; and (4) we know that lung surfactant synthesis is regulated by cell–cell communications, which is consistent with the central concept of this book—that cell–cell interactions are the mechanistic basis for evolution.

One surfactant regulatory gene in particular epitomizes this approach to the development, phylogeny, and evolution of the lung, and meets all four of the criteria, listed above, namely, parathyroid hormone–related protein (PTHrP), which has the following properties:

1. It is a stretch-regulated gene.
2. It has been shown to be necessary for the formation of alveoli, or terminal airsacs, which are the structural and functional means of evolutionary adaptation to atmospheric oxygen for metabolic drive.
3. It is highly conserved in evolution, as it is phylogenetically expressed all the way back to yeast.
4. It signals for surfactant synthesis through cell–cell communications mediated by integrated ligand–receptor mechanisms.

The evolutionary amplification of PTHrP signaling may well be the proximate cause of the developmental and phylogenetic modeling and regeneration of the alveolar wall.

MECHANISM OF MAMMALIAN LUNG DEVELOPMENT

At this point, it would be helpful to review the processes of mammalian lung development, since this book relies heavily on these mechanisms for its insights and understanding of lung evolution. Lung development is divided into two major phases: the branching of the airways, followed by the formation of the alveoli. The lung begins as an outpouching of the primitive foregut, referred to as the *laryngotracheal groove*, at 4–6 weeks of gestation (term gestation = 40 weeks) in humans. The proximal portion of this early structure generates the larynx and trachea, while cells located at the distal end of the trachea give rise to the left and right main stem bronchi. Branching morphogenesis of the left and right bronchi forms specific lobar, segmental, and lobular branches. This process extends through the canalicular stage of lung development up to midgestation. Alveolarization starts at midgestation, continuing on for 8–10 years to generate the 300 million alveoli of the mature lung, the structures where gas exchange occurs. This process generates a gas exchange surface area of 160 m^2, which is the size of a (singles) tennis court, and is 80 times larger than the surface area of the skin, the organ of gas exchange in simpler organisms. This enormous gas exchange surface area is paired with an equally large and efficient alveolar capillary network. The development of the lung is mediated by coordinately integrated, mutually regulated networks of transcription factors, growth factors, matrix components, and physical forces, which all play important roles in determining lung structure and function. During development *in utero*, the lung is filled with fluid. Secretion of alveolar fluid into the airway lumen is osmotically driven by active chloride secretion through its ion channels. This generates the flow of lung liquid into the amniotic sac surrounding the fetus. The larynx acts like a valve, maintaining a positive fluid pressure of approximately 1–2 cm of water in the developing airways. The functional significance of this positive pressure for lung development is revealed by the effect of experimental fluid drainage during fetal life, which results in lung hypoplasia. Conversely, tying off the trachea causes a doubling in the rate of airway branching. Physiologic fluctuations in lung intraluminal pressure due to peristaltic contractions of airway smooth muscle also play an important role in embryonic lung branching morphogenesis. Fetal breathing movements also cause cyclic fluctuations in lung fluid pressure during fetal life.

The discovery that PTHrP is necessary for alveolarization, the primary vertebrate evolutionary strategy for the transition from water to land, and to air breathing, was made serendipitously while studying the effect of PTHrP gene deletion on bone development in mice. Mice that are born missing the PTHrP gene die of pulmonary insufficiency within a few hours of birth. On histologic examination, it was found that these animals had no alveoli. These observations, in combination with the knowledge that PTHrP and its receptor are highly conserved, are stretch-regulated, and form a cell–cell signaling pathway functionally linking the epithelial and fibroblastic germ layers of the embryonic lung to the blood vessels, prompted us to investigate its role in lung phylogeny and evolution. Figure 3.1 represents a broad-view hypothesis, which predicts that for vertebrates, progressively increasing metabolic rates (which are elevated for warm-blooded and active, or cold-blooded species at high temperature) will lead to increased lung complexity [defined as increased surface area, decreased alveolar wall thickness, and increased activity of lung surfactant, as indicated on the y axis (ordinate)]. Such changes are attributed to an increase in PTHrP signaling intensity, as reflected by the x axis (abscissa). Here we explore this hypothesis in further detail.

The key functional feature of PTHrP signaling is that by coordinately stimulating the activities of both PTHrP and its receptor in the endoderm and mesoderm, respectively, alveolar wall distension coordinately increases both surfactant production and alveolar capillary bloodflow—the lung physiologic process referred to as *ventilation/perfusion matching*.

Ventilation/perfusion (V/Q) matching is the net result of the evolutionary integration of cell–molecule interactions by which the lung parenchyma and vasculature have functionally adapted to the progressive increase in the metabolic demand for oxygen—what Krogh called "the call for oxygen." The diffusing capacity of the blood–gas barrier for oxygen exchange correlates directly with the surface area, and inversely with the thickness of the partition. Both ontogenetic and phylogenetic structural adaptation for the increased efficiency of gas exchange is characterized by several features: (1) progressive thinning of the alveolar wall; (2) the concomitant decrease in alveolar diameter; (3) the increased expression of the connective tissue protein type IV collagen, the strongest of all the collagens; and (4) the maximal increase in total surface area. We hypothesize that these structural adaptations could have directly resulted from selection pressure for the phylogenetic amplification of the PTHrP signaling pathway by functionally linking the embryonic endodermal and mesodermal layers, as follows:

1. PTHrP signaling through its receptor is coordinately stimulated by stretching the alveolar parenchyma.

2. Binding of PTHrP to its receptor activates the cyclic AMP-dependent protein kinase A signaling pathway.

3. Stimulation of the PTHrP signaling pathway results in differentiation of the mesodermal connective tissue cells of the alveolar wall, characterized by increased expression of the specific nuclear transcription factor peroxisome proliferator–activated receptor gamma (PPARγ), and its downstream signaling targets adipose differentiation–related protein (ADRP) and leptin.

4. ADRP is necessary for the uptake, storage and trafficking of neutral lipid for surfactant production, and leptin stimulates the differentiation of the alveolar type II cell for surfactant phospholipid and protein production.

5. PTHrP could have affected the cellular composition of the alveoli in a manner consistent with the evolution of the lung in at least three ways (see Fig. 3.2):

    (a) it inhibits lung fibroblast growth, and can cause fibroblast apoptosis, or programmed cell death, which would account for the thinning of the alveolar wall;

    (b) the lung connective tissue stimulation of epithelial type II cell surfactant production by leptin would have compensated for the progressive decrease in alveolar diameter (and the concomitant increase in surface tension), providing the increased surface area for gas exchange; and

    (c) leptin signaling to the epithelial type II cell may also have stimulated type IV collagen synthesis, maximizing cell wall strength without increasing its thickness, since type IV collagen has the highest tensile strength of all of the collagen family members.

The coordinated effects of PTHrP on surfactant production, alveolar capillary bloodflow, and type IV collagen would have facilitated natural selection for the progressive decreases in both alveolar diameter and alveolar wall thickness through ontogeny and phylogeny, increasing the gas exchange surface area/blood volume ratio of the lung, both developmentally and phylogenetically. Mechanistically, PTHrP inhibits myofibroblast (muscle-like fibroblast) differentiation by inhibiting the glioma-associated oncogene, or Gli gene, the second step in the mesodermal wingless/int (Wnt) pathway that determines the myofibroblast phenotype. As a result, PTHrP induces the fat cell–like mesodermal lipofibroblast by upregulating PPARγ, which stimulates neutral lipid trafficking, facilitating surfactant production by alveolar epithelial type II cells. The combined inhibitory effects of PTHrP on both fibroblast and type II cell growth, in concert with PTHrP augmentation of surfactant production, would have had the net effects of simultaneously thinning and buttressing the alveolar wall, promoting selection pressure for the upregulation of PTHrP by physiologically stabilizing what otherwise would have resulted in alveolar collapse, since, according to the law of Laplace, surface tension is inversely

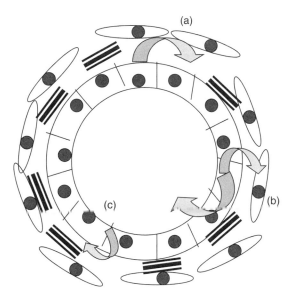

Figure 3.2. Effects of PTHrP on the evolving alveolus: (a) inhibition of fibroblast growth; (b) stimulation of surfactant; (c) stimulation of type IV collagen. (See insert for color representation.)

related to the diameter of a sphere; that is, selection for small alveoli required the evolution of these structural and functional changes, or face extinction.

Evidence that PTHrP regulates lung growth comes from the study of three fundamentally different types of lung: the single-chambered lungs of fish, frogs, and lizards; the bronchoalveolar lung of mammals; and the cross-current saccular lung of birds. Frog and lizard terminal airsacs, or *faveoli*, are up to 100 times larger in diameter than the alveoli of a similarly sized mammal. Moreover, although the epithelial lining cells can be distinguished as type I and type II cells, the type II cells are not as well delineated as those in the alveoli of mammals. In addition, the alveolar wall is heavily populated with myofibroblasts, the muscle-like connective tissue cells. The very large diameter of the frog faveolus exempts it from the constraints of surface tension. In addition, the faveolar interstitial muscle cells facilitate its expansion and contraction for gas exchange.

In mammals, the conducting and terminal airways of the lung are convoluted, promoting gas exchange in association with increasing metabolic demand for oxygen. The structural and functional changes seen during phylogeny may be explained by the increased amplification of PTHrP signaling through different levels of distension of the alveolar wall in different phyla, as depicted in Figure 3.2. PTHrP signaling inhibits myofibroblast differentiation by downregulating the Wnt pathway, thus potentially explaining the tandem trends toward the decreasing numbers of myofibroblasts in the lungs in association with smaller alveoli and thinner alveolar walls, leading to an increase in the efficiency of gas exchange. For example, Daniels and Orgeig (2003) have demonstrated that the amount of pulmonary surfactant per unit surface area declines, but the level of saturation of the phospholipid fatty acids increases between frogs and reptiles on one hand and among mammals on the other hand. It is likely that the increased saturation and the greater turnover of surfactant in mammals has acted to sustain the structural integrity of the smaller alveoli.

AVIAN LUNG STRUCTURE–FUNCTION RELATIONSHIP: THE EXCEPTION THAT PROVES THE RULE

Birds are an exception in vertebrate phylogeny because their lungs lack alveoli; the bird lung is stiff and is firmly attached to the dorsal thorax, preventing the expansion and contraction of the lung seen in amphibians, reptiles, and mammals. Yet the bird lung possesses epithelial type II cells that express PTHrP and surfactant; in birds, the surfactant system is not necessary for the prevention of alveolar collapse due to surface tension, and may, as in amphibians, be required only for host defense or for preventing the leakage of blood from the vasculature, which is highly susceptible to injury because it protrudes far into the airspace. Birds may have evolved "overengineered" lungs in response to the selection advantage of efficiently oxygenating in flight. Bird lungs are ≤4 times more efficient in exchanging oxygen and carbon dioxide with the atmosphere as mammals. We speculate that because PTHrP is necessary for bone morphogenesis, and the hollow bones of birds that enable them to fly are connected to their respiratory system (because they contain part of the airsac system), control of development of the lung and skeletal phenotypes may be tightly evolutionarily interlinked. It is noteworthy that the chick PTHrP signaling mechanism for surfactant is relatively weak (Torday and Rehan, unpublished observation) compared to that of mammals, suggesting a convergent molecular adaptation for decreased PTHrP signaling in both the lung and bone of avians, possibly as a coevolved evolutionary strategy for flight. To paraphrase the old chestnut about which came first, the chicken or the egg, a systematic study of PTHrP signaling in the lung and bone of flightless versus flying birds might provide the answer to this question.

DOES ONTOGENY RECAPITULATE PHYLOGENY? THE ROLE OF PTHrP IN LUNG DEVELOPMENT

The PTHrP gene is essential for mammalian lung development—deletion of this gene results in failed alveolization, which is also phylogenetically relevant because the progressive generation of alveoli is the strategy by which the lung has evolved in vertebrates. In mice lacking the PTHrP gene, this deletion results in death due to pulmonary insufficiency within minutes to hours after birth. Structurally, the PTHrP-deficient lung develops up to the canalicular stage, coincident with the point in lung development when PTHrP is produced in the mouse, rat, and human. Functionally, the lung does not produce surfactant, consistent with failed alveolization due to immaturity of the embryonic lung tissue layers. Phylogenetically, PTHrP is expressed in the swim bladder of fish, as is surfactant. Both PTHrP and surfactant are expressed in the lung of every vertebrate examined so far. In frogs, the PTHrP receptor has been profiled during metamorphosis. It is upregulated in the tadpole prior to air breathing in association with the appearance of the lungs. Bird lungs also express PTHrP, and the profile of PTHrP expression parallels that of surfactant content, PTHrP mRNA increasing one day prior to the accelerated increase in surfactant phospholipid production in chickens, suggesting a causal relationship between amplification of PTHrP signaling and lung development. However, the chick lung does not have alveoli; it has airsacs. Because either overdistension or constant (vs. cyclic) distension of the alveolar type II cell causes downregulation of PTHrP expression in rodents, we speculate that birds do not alveolarize because of their stiff, nonreciprocating, flowthrough lungs, which are 4 times more efficient in exchanging gases than the most efficient mammalian lung, taking maximal advantage of their jet-engine-like design for optimal gas exchange during flight.

INTERRELATIONSHIP BETWEEN PTHrP, DEVELOPMENT, PHYSIOLOGY, AND REPAIR: IS REPAIR A RECAPITULATION OF ONTOGENY AND PHYLOGENY?

The processes that occur during mammalian lung development in the fluid-filled womb prepare the fetus for birth and physiologic balance, or homeostasis. The development and maturation of the lung is key to the transition to postnatal air breathing, because surfactant production is necessary for effective gas exchange. On the basis of this functional linkage between lung development and homeostasis, we have generated data demonstrating that the underlying mechanisms of repair may recapitulate ontogeny. If lung fibroblasts are experimentally deprived of PTHrP, which occurs when the lung alveolar epithelial type II cell is injured, their structure changes, as follows. The first function that decreases is the PTHrP receptor on the interstitial fibroblast surface, which is progressively lost over time, as are its downstream signaling targets, such as PPARγ, adipocyte differentiation–related protein (ADRP), and leptin; the decline in the PTHrP-dependent phenotype of the lung fibroblast is mirrored by the stepwise gain of the muscle-like phenotype, characteristic of fibrosis. This structural change can be worsened by transforming growth factor β_1, or *reversed* by treatment with PTHrP. The readaptation of the fibroblast in response to the loss of the epithelial–mesenchymal PTHrP–leptin crosstalk is indicative of the balance between homeostasis and repair, particularly the early loss of the cell surface PTHrP receptor as a functional determinant of this cell type. We speculate that lipofibroblast repair recapitulates ontogeny because it was programmed to express the developmental crosstalk through evolutionary selection pressure for homeostasis. In the model that has been developed to test this hypothesis, there are three key principles:

1. The crosstalk between epithelium and mesoderm is necessary for the maintenance of homeostasis.
2. Damage or injury to the epithelium muffles or silences the crosstalk, leading to loss of homeostasis/readaptation to a newly established homeostatic setpoint (i.e., myofibroblast proliferation).
3. Depending on the nature and severity of the injury (mild, moderate, or severe; proximal or distal to the regulatory mechanisms for homeostasis, etc.), and the condition of the host (healthy, compromised, polymorphisms for key genes necessary for reestablishing homeostasis, infected, premature, etc.), either (a) normal physiology will be regenerated; (b) cell and tissue remodeling may occur, causing altered lung function; and/or (c) fibrosis will persist, leading to chronic lung disease.

Cell–molecular injury affecting epithelial–mesenchymal crosstalk recapitulates ontogeny and phylogeny, or evolution, providing novel and biologically effective diagnostic and therapeutic molecular targets. On the basis of the more general concept that phylogeny provides stable phenotypes for evolution and ontogeny, we speculate that PTHrP signaling in the lung parenchyma of fish, frogs, reptiles, mammals, and birds recapitulates the *cis* regulatory amplification of the surfactant mechanism, providing the means for the alveoli to decrease in diameter, increasing the surface area/blood volume ratio, facilitating gas exchange with the circulation.

In the current chapter, we have further expanded on the concept of cell–cell communication as the basis for the evolution of the lung. The lung is prototypical for the evolution of vertebrate physiology from uni- to multicellular organisms. In Chapter 4 we will build on the cell–cell communication mechanism as the basis for physiologic evolution using developmental biology as a way of linking the lung platform to other tissues, organs, and physiologic principles.

REFERENCES

Bloch KE (1979), Speculations on the evolution of sterol structure and function. *CRC Crit. Rev. Biochem.* 7(1):1–5.

Daniels CB, Orgeig S (2003), Pulmonary surfactant: the key to the evolution of air breathing. *News Physiol. Sci.* 18:151–157.

Darwin C (1859), *On the Origin of Species by Means of Natural Selection, or the Preservation of Favoured Races in the Struggle for Life.* John Murray, London.

Torday JS, Rehan VK (2004), Deconvoluting lung evolution using functional/comparative genomics. *Am. J. Respir. Cell. Mol. Biol.* 31(1):8–12.

4

THE EVOLUTION OF CELL–CELL COMMUNICATION

Two roads diverged in a wood, and I—
I took the one less traveled by,
And that has made all the difference.
—*Robert Frost, The Road Not Taken*

Evolutionary Biology, Cell–Cell Communication, and Complex Disease, First Edition.
John S. Torday and Virender K. Rehan.
© 2012 Wiley-Blackwell. Published 2012 by John Wiley & Sons, Inc.

As a preamble to this chapter, we present the suggestion by E. O. Wilson in his book *Consilience* that since all human knowledge is rapidly being reduced to 1s and 0s by computers, we could create a common database across all disciplines. In so doing, Wilson challenges us to generate a unifying theory for biology in order to fulfill this promise, because without such a theory, we have only disorganized information and anecdotes (Wilson 1998).

CELL–CELL COMMUNICATION AS THE MECHANISTIC BASIS FOR EVOLUTIONARY BIOLOGY

Cell–cell communication provides common ground for understanding complex biologic processes ranging from development to physiologic homeostasis, regeneration, and aging, all of which are functional phenotypes generated by evolution. But more importantly, *cell–cell communication is the mechanistic basis for the process of evolution*. Evolution can be reduced to three forms of communication: (1) at the origin of vertebrate evolution, between unicellular organisms and their physical environment; (2) the cell–cell communications that form the basis for multicellular physiology; and (3) the communication of genetic material from one generation to the next, or reproduction. This conceptualization of evolution should not be confused with the writings of Jablonka and Lamb (2006), or of Maynard Smith and Eors Szathmáry (1995), who focus on biologic information per se to understand evolution, rather than on the evolved processes of cell–cell communication. This is a subtle but important distinction, because it distinguishes between reduction to the components of a process, as opposed to focusing on the process itself.

There have been many attempts to conceptualize integrated biology, beginning with the great chain of being (Linnaeus' binomial nomenclature), and Darwin's *On the Origin of Species*. More recently, however, it seems that the closer we get to the basic elements of biology, the more skepticism we experience regarding whether we can or will ever determine the central dogma of biology. This is predictable, since we are getting into uncharted territory, exceeding the comfort level for those who are wed to the old descriptive paradigm. Molecular biologists have rediscovered the gap between the genotype and the phenotype yet again, but are unable to connect them. Witness the essay "Life's irreducible structure" by Polanyi (1968), who argued that the form and function of the various parts of living organisms *cannot* be reduced to (or explained in terms of) the laws of physics and chemistry, and so life exhibits what he referred to as *irreducible structure*. He did not speculate on the origins of life, arguing only that scientists should be willing to recognize the impossible when they see it. Molecular biology has discovered that in the transition from genes to phenotypes there are multiple levels of control. Each level of control is determined by a system of self-regulating proteins that operate by the rules and laws of what we term a *data operating system* that we do not yet understand. Polanyi expressed this phenomenon using the game of chess as a metaphor. Each move that the chess player makes is constrained by the boundaries of the game, but we do not know what those boundaries are. In contemporary molecular biology, we describe the equivalent of chess moves made by genes and proteins, ignoring the overall laws of the data operating system that govern those genes and proteins. As a result, we are unable to integrate the environmental signals with the phenotypic patterns.

At this juncture, we mention Prigogine's assessment of life's irreducible complexity in his book *Order Out of Chaos* (Prigogine 1984), in which he concludes that biology is too complicated to define. In contrast to such attempts to understand biology by analyzing it in its present form, in this book we have approached the question of the data operating system underlying evolution by starting from its cellular origins, moving forward in biologic time. This is analogous to the physicists viewing the universe as having originated from the big bang, and understanding such phenomena as the patterned distribution of the elements and the cosmic microwave background, with the formation of black holes and supernovas as a result.

In the midst of the sea change that we are undergoing in the postgenomic era, it is helpful to step back and reassess the situation in order to gain perspective on the processes biology and medicine. The major take-home message of the Human Genome Project is that humans have fewer genes than a carrot does (25,000 vs. 40,000), whereas it had been predicted that we would have at least 100,000 genes, based on the relative biologic complexity and number of genes found in worms, flies, and humans—so much for prediction based on descriptive biology. The fact that humans have fewer genes than expected does not mean that we are "simpler" than organisms with more genes—such logic derives from *ex post facto* reasoning and descriptive biology. It is more likely that we have used ancestral genes more effectively as a consequence of the evolutionary process. Moreover, although we do not yet know what the specific mechanisms of evolution are, we have gained some insights through a developmental cellular–molecular approach to the process of lung evolution.

THE DARWINIAN BIOLOGIC SPACETIME CONTINUUM AND EINSTEIN'S VISION OF THE UNIVERSE

During his voyage on *HMS Beagle*, Darwin saw a continuum of species generated by natural selection, unlike the anthropocentric great chain of being, which was the prevailing way of thinking about all God's creatures at the time. However, Darwin's explanation for the biologic patterns he observed was survival of the fittest, which is only a metaphor for the evolutionary process, not a mechanism. This was a radical change in the way humans perceived their evolutionary origins. In the interim, however, our fundamental knowledge of biology at the cell–molecule level has demanded a mechanistic and predictive understanding of evolutionary biology. After all, it has been accomplished in physics and chemistry, from which biology is derived. Also, a working model of evolutionary biology is essential if we are going to take full advantage of the human genome, and the genomes of model organisms. For example, the cellular–molecular mechanism for lung evolution depicted in the schematic in Figure 4.1 infers that there is a continuum from development to homeostasis and repair. The depiction of the process of lung evolution infers that there is a direction and magnitude of change. This perspective is similar to Einstein's vision, as a 16-year-old, of traveling through space in parallel with a light beam (see Fig. 4.2), which seems to have sparked his revolutionary insight to the continuum from Brownian movement to the photoelectric effect and relativity theory, as reflected by his publication of groundbreaking scientific papers on each of these subjects in rapid succession in 1905, his so-called miracle year (wunderjahr).

Since our model of evolution is based on universally accepted developmental biologic principles, the evolution of all the other tissues and organs is amenable to the same mechanistic approach.

REVERSE ENGINEERING OF PHYSIOLOGIC TRAITS AS A PORTAL FOR VIEWING EVOLUTION

The inquiry into the mechanisms and laws of nature is traditionally either from the top down or from the bottom up. The *top–down approach* begins with an observation for which the observer formulates the components, how they function, and a mathematical expression to interrelate them. A series of experiments is then designed to support or refute the model, similar to Michaelson and Morley's experiment to determine whether the "ether" through which light was thought to travel existed. In the *bottom–up approach*, the observer collects data related to the phenomenon in question, characterizes them under various controlled conditions, and devises experiments to further explore the observed data and their properties. As a result, the observer formulates a working model for further testing. The top–down approach is commonly used in physics, which lends itself to mathematical modeling, whereas the bottom–up approach traditionally used in biology is context-dependent, and therefore anecdotal. This fundamental difference between physics and biology may be due to the fact that biology is derived from physics through an undetermined mechanism, euphemistically referred to as *natural selection*. A third strategy is from the *middle–out approach*, which links mechanisms

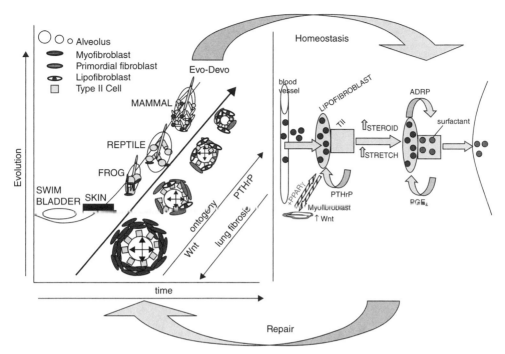

Figure 4.1. Lung biologic continuum from ontogeny–phylogeny to homeostasis and repair. The schematic compares the cellular–molecular progression of lung evolution from the fish swim bladder to the mammalian lung (left portion) with the development of the mammalian lung, or evo-devo, as the alveoli become progressively smaller (see legend in upper left corner), increasing the surface area-blood volume ratio. This is facilitated by the decrease in alveolar myofibroblasts, and the increase in lipofibroblasts, due to the decrease in Wnt signaling, and increase in PTHrP signaling, respectively. Lung fibrosis progresses in the reverse direction (lower left corner). Lung homeostasis (right portion) is characterized by PTHrP/leptin signaling between the type II cell and lipofibroblast, coordinately regulating the stretch regulation of surfactant production and alveolar capillary perfusion. Failure of PTHrP signaling causes increased Wnt signaling, decreased PPARγ expression by lipofibroblasts, and transdifferentiation to myofibroblasts, causing lung fibrosis. Repair (arrow from homeostasis back to ontogeny–phylogeny), is the reca-pitulation of ontogeny–phylogeny, resulting in increased PPARγ expression. (See insert for color representation.)

to structures. This is the premise for the approach that we have taken to evolution. By tracing the ligand–receptor-mediated cell–cell communications that have determined the pulmonary surfactant phenotype backward in time and space developmentally, both within and between species, we can reverse-engineer the cellular–molecular mechanisms that have fashioned the lung through selection pressure. We refer to this as a *middle–out approach*, in contrast to the top–down and bottom–up strategies.

Horowitz formulated a similar reverse-engineering approach to the evolution of biochemical pathways back in 1945. He postulated a retrograde mode of

Figure 4.2. Einstein's wonder year. In 1905, Einstein published three groundbreaking papers on Brownian movement, the photoelectric effect, and the theory of relativity. He described having an epiphany at age 16, picturing himself traveling in parallel with a beam of light. That vicarious experience may have given him the insight for these papers.

evolution, as follows. He envisioned an organism that could not synthesize a biochemical substance essential for life, forcing it to acquire the substance from its environment, or become extinct. When the supply of that substance in the environment was exhausted (due to the reproductive success of the organism), those organisms that possessed the last enzyme in the biosynthetic pathway could make use of the immediate precursor, converting it to the end product, until the supply of the immediate precursor was also exhausted. Then only those organisms that possessed the next-to-last (penultimate) enzyme could survive, and so on and so forth, until the complete biosynthetic pathway was ultimately in place.

This approach ingeniously describes the evolutionary process from the present backward, but does not account for how selection pressure determines the biologic cell–cell communication mechanisms that allowed it to form. By contrast, the cellular–molecular, middle–out approach provides the developmental and phylogenetic cytoarchitectural, ligand–receptor-mediated mechanisms that gave rise to the biosynthetic pathway. Selection pressure for such ligand–receptor gene regulatory networks generated both evolutionary homeostatic stability and novelty through mechanisms such as gene duplication, gene mutation, redundancy, alternative pathways, compensatory mechanisms, positive and balancing selection pressures, alternative splicing, and *cis* regulation. Such genetic modifications were manifested by the structural and functional changes in the blood–gas barrier, primarily the thinning of the blood–gas barrier for gas exchange, in combination with progressive, adaptive phylogenetic changes in the composition of the surfactant, as described by Daniels and Orgeig (2003). The reverse-engineering of these phenotypic changes in the blood–gas barrier forms the basis for the molecular genetic approach to lung evolution put forward in this book.

CELL–CELL COMMUNICATION AS THE BASIS FOR THE EVOLUTION OF METAZOANS

On the basis of the middle–out, cell–cell communication model, how could biologic evolution have begun? As indicated in Chapter 1, one school of thought is that cellular organisms first emerged as lipids derived from algae, which are 90% lipid, as they are deposited on the shores of lakes, rivers, and oceans. The alternate wetting and drying of these lipids by the water and sun would have generated *micelles*, which are semipermeable membranes that could have provided a protected environment for the reduction of entropy through enzymatic catalysis (see Chapter 1, Fig. 1.1, structure **V**). Over the next 4.5 billion years, unicellular organisms evolved in adaptation to their physical surroundings. Eukaryotes evolved from prokaryotes; the former were distinguished from the latter primarily by the presence of a nuclear envelope. That perinuclear membrane may well have evolved to protect the eukaryotic nucleus against invasion by prokaryotes—and the eukaryotic acquisition of mitochondria from prokaryotes, termed the *endosymbiosis theory*, similarly emerged as a result of the ongoing competition between pro- and eukaryotes. Then, about 500 million years ago, unicellular organisms began cooperating with one another metabolically. In the case of the prokaryotes, this took the form of such phenomena as biofilm and quorum sensing. Such cell-cooperative adaptations may have been matched by the selection pressure for eukaryotes to metabolically cooperate (or become extinct), resulting in evolution of the cell–cell signaling mechanisms that we recognize today as the soluble growth factors that mediate morphogenesis, homeostasis, regeneration, and reproduction.

However, it should be borne in mind that this process began with a decrease in entropy, which violated the second law of thermodynamics, with unicellular organisms fending against physical (gravity, oxygen, etc.) forces, eukaryotes fighting off prokaryotes, and in the process, developing the mechanisms of homeostasis and physiology. But the laws of physics cannot be broken, so there had to be some form of energy payback. Unicellular organisms are immortalized by reproduction through binary fission, whereas eukaryotes evolved asexual, and then sexual, reproduction as a means of communicating their genetic information from one generation to the next. If the ultimate evolutionary selection pressure is for reproductive success, then the energy distribution is not symmetric throughout the lifecycle, as in unicellular organisms reproducing by binary fission, but instead favors the reproductive phase of the lifecycle (see Fig. 1.1, **V**). The tradeoff is that the cellular machinery must ultimately fail because the bacteria, the oxidative stress, and gravitational forces that initiated the evolutionary strategy are omnipresent. The net result is decreased *bioenergetics* in the postreproductive stage of the life-cycle (see Fig. 1.1, **V**), resulting in such well-recognized phenomena associated with the aging process as increased oxidative stress, lipid peroxidation, protein misfolding, endoplasmic reticulum stress, breakdown in cell–cell communication due to decreased receptors, and failure of other such metabolic mechanisms. All of these factors have been implicated in the process of aging, without an integrated mechanism to explain how or why they evolved, with concepts such as antagonistic pleiotropy or the disposable soma theory merely describing the hallmarks of the process, but not providing a mechanism. By contrast, the bioenergetic imbalance caused by evolutionary selection pressure favoring reproduction causes decreased cell–cell communication (as is the case for PTHrP signaling, for example) because it is an energy-requiring process, ultimately resulting in a breakdown in homeostasis (and the predicted loss of alveoli, as has been shown in aging rats), culminating in death. But the organism's gene pool is, in effect, immortalized by communicating its DNA from one generation to the next through reproduction.

In the final analysis, each aspect of this perspective on the how and why of evolution is reducible to cell–cell communication, initially between unicellular organisms and their physical environment, followed by cell–cell communication as the basis for metazoan structure and function, and finally through reproduction as the communication of genetic information from one generation to the next.

The cellular cooperativity that underlies the emergence of eukaryotes has evolved from metabolic processes to cellular forms that have been recapitulated throughout the evolution of multicellular organisms as phylogeny and ontogeny. Consider, for example, the epithelial–mesenchymal interactions that form tissues and organs. Such interactions are necessary for the formation of the liver, as well as its homeostatic control of lipids, which shuttle back and forth between the systemic circulation, the hepatic stellate cells, and neighboring hepatocytes. Epithelial–mesenchymal cell–cell interactions also determine the development and regulation of organs as diverse as the gut, pancreas, kidney, eye, and endocrine tissues such as the adrenals, gonads, prostate, and mammary gland.

UNDERSTANDING LUNG EVOLUTION FROM THE MIDDLE OUT

The challenge for biology and medicine in the postgenomic era is to integrate functional genomic data in order to determine the first principles of physiology. Currently, this problem is being addressed statistically, by analyzing large datasets to identify genes that are *associated* with structural and functional phenotypes—

whether they are actually causal is largely ignored. This approach is just an extrapolation of the *Systema Naturae* published by Linnaeus in 1735. The reductionist genetic approach cannot simply be used to compute phenotypes. Evolution is not a result of chance; it is an *emergent and contingent* process. Ironically, the cosmologist Lee Smolin (1997) has applied Darwinian selection to stellar evolution, hypothesizing that there is a mechanistic continuum from elementary particles to the formation of black holes, while biologists still don't have a working model for evolutionary biology.

In the current and future research environments, we must expand our computational models to encompass a broad evolutionary approach, because as Dobzhansky (1973) cleverly put it, "Nothing in biology makes sense except in the light of evolution," but we have yet to achieve this vision. We have formally proposed using a comparative, functional genomic, middle–out approach to solve for the evolution of physiologic traits. This approach engenders development, homeostasis, and regeneration as a family of parallel lines that can be mathematically analyzed as a set of simultaneous equations (see Chapter 3, Fig. 3.1), from which we can derive the overall processes of evolution. This perspective provides a *feasible and refutable* way of systematically integrating such information in its most robust form to retrace its evolutionary origins (see Figure 4.2). This is referred to as a diachronic analysis of a phenomenon across time, in contrast to the traditional synchronic analysis of evolution at a single point in time, namely, the present.

THE CELL–CELL COMMUNICATION MODEL OF LUNG EVOLUTION TRACES CONTEMPORARY PHENOTYPES BACK TO ANCESTRAL PHENOTYPES

We have extended the developmental and comparative aspects of the leptin effect on mammalian lung development to frog lung development. Crespi and Denver (2006) have shown that leptin stimulates tadpole limb development. This was an interesting observation because it provided a pleiotropic mechanism for the evolution of vertebrates, since metabolism, locomotion, and respiration are the three driving forces behind this process. To test the hypothesis that leptin biology might be a working model for vertebrate evolution, we treated *Xenopus laevis* tadpole lungs with frog leptin, and, surprisingly, found that it has the same effects that it has on mammalian lung—it causes thinning of the blood–gas barrier in combination with increased production of surface-active phospholipid and surfactant proteins (Torday et al. 2009). These effects of leptin on surfactant are paradoxical since the frog lung terminal airspaces, termed *faveoli*, are so large in diameter and muscular that they do not require surface-tension-reducing activity to function. However, leptin has also been shown to stimulate surfactant protein A expression, which is an antimicrobial peptide. This, and the fact that antimicrobial peptides are expressed in the gut and skin, suggest that the original selection pressure may have been for host defense, which was exapted (preadapted) for blood–gas barrier expansion and host defense in the gut, lung, and skin.

This scenario has cast new light on our studies of the effects of bacterial infection on lung development. It is well known that intrauterine infection accelerates lung surfactant production, so we tested the effect of the bacterial wall constituent lipopolysaccharide (LPS) on developing lung epithelial cells. Remarkably, LPS had the same stimulatory effect on lung development as leptin does, suggesting that the original stimulus for the epithelial–mesenchymal interaction may have been due to extrinsic stimulation by invading bacteria, subsequently mimicked evolutionarily by the intrinsic leptin mechanism. These interrelationships may, in turn, relate back to the swim bladder origins of the lung, since the lung is homologous to the swim bladder of physostomous fish, which have a tube connecting the swim bladder to the gut, similar to a trachea. This physical connection also provides access for bacteria to enter the swim bladder from the gut. In contrast, the swim bladders of physoclistous fish have no tracheal connection to the esophagus. In support of this homology, physostomous fish must come to the surface after hatching in order to inflate their swim bladder with air, which is the equivalent of the first breath that mammalian newborns take at birth. In either case, surfactant deficiency results in neonatal death.

PREDICTIVE VALUE OF THE LUNG CELL–CELL COMMUNICATION MODEL FOR UNDERSTANDING THE EVOLUTION OF PHYSIOLOGIC SYSTEMS

Unlike the classic pathophysiologic approach to understanding health, which reasons backward from disease, the evolutionary–developmental approach, as depicted in Figure 4.2, reasons in the forward direction, starting from the cellular origins of physiology, resulting in prediction of the cause of chronic disease, as we have shown for the lung and Leon Fine has shown for the kidney. Fine states that "Regeneration seems to follow the same pattern of sequential differentiation steps as nephrogenesis. The integrity of the epithelium is restored by re-establishing only those stages of differentiation that have been lost. Where cell death occurs, mitogenesis in adjacent cells restores the continuity of the epithelium and the entire sequence of differentiation events is initiated in the newly generated cells" (Bacallao and Fine 1989). The burgeoning literature on stem cells has fostered similar thinking with regard to regeneration of the retina, pancreas, liver, and neurons.

In further support of the middle–out, cell–cell communication approach, we will cite other fundamental differences between a pathophysiologic versus an evolutionary approach to disease. For example, PTHrP is a stretch-regulated gene that mechanistically integrates the inflation and deflation of the alveolar wall with surfactant production and alveolar capillary perfusion. PTHrP is conventionally regarded as a bone-related gene that regulates calcium metabolism. With these two disparate properties of PTHrP in mind, we recalled that astronauts develop osteoporosis as a result of weightlessness in deep space. Lung physiology is also affected by zero gravity (0 g) (West et al. 1997). The normal alveolar ventilation/perfusion (V/P) gradient from the apex to the base of the lung is lost in microgravity, increasing the work of breathing. Since PTHrP integrates ventilation/perfusion matching, this phenomenon may also be due to the influence of microgravity on PTHrP.

Others have pursued the traditional pathophysiologic approach to the phenomenon of microgravity-induced osteoporosis, reasoning that postmenopausal women are vulnerable to development of osteoporosis and are estrogen-deficient, concluding that estrogen treatment would be effective for microgravitational osteoporosis. On the other hand, we have tested the hypothesis that exposure to microgravity would inhibit PTHrP expression in astronaut bone and lung cells. We observed a decrease in PTHrP expression by osteoblasts and lung epithelial cells in freefall attached to buoyant dextran beads, which mimicks 0 g. When these cells were put back in unit gravity (1 g), the expression of PTHrP returned to normal levels in both cell types.

We subsequently examined the PTHrP expression by the bones of rats flown in deep space for 2 weeks. Here, too, we observed a significant decrease in PTHrP expression compared to ground-based control rat littermates. The effect of weightlessness was not seen in the non-weight-bearing skull bones, consistent with the hypothesized effect of microgravitational unloading of the weight-bearing bones.

We have similarly taken an unconventional evo-devo biologic approach to chronic lung disease. We have pursued the concept that there is an evolutionary continuum from development to homeostasis and regeneration mediated by soluble growth factors. On the basis of that approach, we have discovered that the cell–cell communication between the lung epithelium and mesoderm is critically important for the development and maintenance of the alveolar lipofibroblast, and that when that signaling mechanism fails, the lipofibroblast defaults to its developmental cellular origin as a myofibroblast, the signature cell type for fibrosis. This approach has given us insight into not only the complex, multifactorial causes of bronchopulmonary dysplasia (BPD) [pressure, oxygen, infection, maternal smoking (nicotine)] but also a novel treatment for BPD, the chronic lung disease of preterm birth caused by PTHrP deficiency. This approach is based on PTHrP regulation peroxisome proliferator–activated receptor gamma (PPARγ) expression, the nuclear transcription factor that determines the lipofibroblast phenotype. Thiazolidenediones are potent PPARγ agonists, and we have found that they can prevent or reverse the effects of all of the BPD-inducing agents we have studied, ranging from pressure-induced injury, to excess oxygen, infection, and nicotine. This evolutionary–developmental approach may be far more successful in the treatment of BPD than the use of the more traditional antiinflammatory agents, such as steroids, or prophylactic surfactant therapy.

SEXUAL DIMORPHISM OF LUNG DEVELOPMENT: A CASE STUDY IN CELL–CELL COMMUNICATION AND EVOLUTIONARY PLASTICITY

It has long been recognized that lung structure is sexually dimorphic, and the mechanism for this became a focal point back in the mid-1970s because it had been observed among pregnant women treated antenatally with glucocorticoids for lung immaturity that the risk of respiratory distress syndrome (RDS) was halved in females, but had no effect on males. This experiment of nature provided a means of discovering the underlying nature of the cellular machinery that mediates the processes of lung development. The global strategy was to exploit the spontaneous sex difference in fetal lung development that occurs across a variety of mammalian species *in utero*. This approach would allow the teasing out of the clinically significant differences in the timing of fetal lung development representing the small incremental steps that typify self-organizing systems with respect to both ontogeny and homeostatic control to reveal the underlying mechanisms of neonatal lung disease—in other words, the evolutionary strategy for lung physiology.

The first empiric evidence for a link between fetal sex and lung maturation was the observation of a sex-specific pattern of surfactant accumulation in human amniotic fluid. There was a 2-week difference in the late-gestation upsurge of surfactant between males and females, which is equivalent to a 20% difference in the risk of RDS, or surfactant deficiency, at birth. This observation was similar to that of Kotas and Avery (1980) that there was a sex difference in fetal lung structure and surfactant production, reflected by the lung pressure–volume curve, a standard measure of lung compliance, during a comparable phase of fetal lung development in fetal rabbits. Adamson and King (1984) characterized this sex-specific difference in the rate of lung maturation in the fetal rat in a subsequent series of studies of the lung cytoarchitecture and surfactant biology, finding a similar pattern of female lung development preceding that of males. Independent studies by our laboratory demonstrated spontaneous sex differences in surfactant synthesis in both fetal rats and rabbits, respectively. This was followed by the demonstration that the sex difference was due to a variation in surfactant synthesis by cultured fetal rat lung alveolar epithelial type II cells.

A survey of the hormones in fetal circulation that might be responsible for the sex difference in fetal lung development suggested three candidates: androgens, müllerian inhibiting substance, and follicle-stimulating hormone. Experimental evidence confirmed that the difference in the rates of lung development was due to the effects of androgens. The physiologic effect of androgens on peripheral tissues is dependent on local activation of testosterone by 5α-reductase, followed by binding to the androgen receptor. Nielsen (1985) had demonstrated the dependence of the sex difference on the androgen receptor using the mouse model of testicular feminization (Tfm), which is devoid of androgen receptors. As in the rat and rabbit, there is a spontaneous sex difference in the amount of androgen receptor; androgen receptor activity has been demonstrated in human lung fibroblasts.

Experiments in our laboratory demonstrated that the androgen receptor inhibitor flutamide eliminated the sex difference in surfactant production by fetal rabbits by neutralizing the inhibitory effect of androgen on lung surfactant production. In initial studies of the mechanism of androgen action on fetal lung development we had observed that androgen blocked the production of fibroblast pneumonocyte factor (FPF), a low-molecular-weight protein produced by lung fibroblasts that stimulates surfactant synthesis by type II epithelial cells, but had no effect on the type II epithelial cell response to FPF, indicating that the primary cellular site of androgen action is on the developing fetal lung fibroblast. These experimental results were consistent with the observation that there was a constitutive difference in the ability of male and female fetal lung fibroblasts to produce FPF in culture, but that the type II cells were able to respond to FPF independently of the sex of the donor cells. To test the hypothesized effect of androgen on the fibroblast glucocorticoid response mechanism, fetal rats were treated with androgen *in utero*, and the fetal rat lung tissue was assayed for both glucocorticoid receptor and 11-oxidoreductase activities, both of which physiologically mediate the effects of cortisol on lung surfactant; androgen delayed the onset of both mechanisms. Fibroblasts were then isolated from the fetal lung tissue, and we observed the same inhibitory effects on glucocorticoid receptor and 11-oxidoreductase activities by cultured fibroblasts. Furthermore, androgen inhibits FPF production and expression by fetal rat lung fibroblasts, probably by inhibiting these glucocorticoid-responsive mechanisms, since the androgen-treated fibroblasts were also unresponsive to cortisol treatment in culture. In support of this mechanism of androgen action, Sweezey et al. (1998) have shown that androgen inhibits the glucocorticoid receptor in fetal rat lung.

Since it had been observed that both glucocorticoids and androgens could alter the phenotype of developing lung fibroblasts (e.g., FPF production, 11-oxidoreductase, glucocorticoid receptor), we performed experiments to determine whether immature fibroblasts constitutively express inhibitory factors that might mediate the androgen effect on lung epithelial type II cell maturation. It was observed that secretions from immature fetal rat lung fibroblasts inhibited type II cell surfactant stimulatory bioactivity from mature lung fibroblasts. This inhibitory activity was neutralized by an antibody to transforming growth factor beta (TGFβ); moreover, TGFβ mimicked the inhibitory effect of the endogenous factor on FPF, and immature fetal rat lung fibroblasts produced TGFβ. Androgen treatment maintained the production of TGFβ by fetal rat lung fibroblasts;

conversely, glucocorticoid treatment *in vivo* caused a decrease in the production of TGFβ by these cells.

ANDROGEN AFFECTS THE EXPRESSION OF GROWTH FACTORS INVOLVED IN LUNG DEVELOPMENT

Exposure of fetal rabbits to a continuous infusion of androgen beginning at the time of sexual differentiation inhibits the increase in lung epidermal growth factor receptor (EGFR) density, and inhibits EGFR activity, while increasing TGFβ activity in fetal mice.

There is also extensive evidence that there is a genetic basis for sex-specific mortality during embryonic development. For example, we have observed that chick lung maturation *in ovo* is delayed in females, suggesting that this mechanism is not so much male versus female as it is a function of the homogametic versus the heterogametic sex in nature. A biologic relationship between homo- and heterogametic sex and perinatal viability is supported by the observations that among mammals it is the males who are at increased risk in the newborn period, whereas among birds it is the females. These genetically based phenotypic characteristics of sexually dimorphic fetal lung development point to an evolutionary mechanism, since sexual reproduction evolved from asexual reproduction. In the process of converting from a single-sex phenotype to two sexes during intrauterine development, all conceptuses begin as the phenotype of the homogametic sex (XX), and the genetically determined males (XY) must be actively transformed to the phenotype of the heterogametic sex by producing androgens. The converse is true in birds—all embryos begin life as the homogametic (ZZ) male phenotype, and will be transformed into females if they are genetically heterogametic (ZW). On the basis of these observations, testosterone appears to be used evolutionarily to eliminate undesirable androgen-related genetic characteristics in the population. Similarly, the absence of a second allele on the Y chromosome places male embryos at greater risk of demise if a deleterious mutation or deletion of the single-gene copy on the X chromosome occurs. Quantitatively, the number of genetic stillbirths in the first trimester is much greater than the number of males that die as a result of the androgen effect on lung development in the third trimester. However, the biologic investment is much greater in the third trimester; furthermore, the androgen will also increase fetal size, putting the mother at increased risk of death during childbirth due to disproportion between fetal head circumference and the maternal pelvis size; loss of the mother is the highest price that the species pays regarding contribution to the gene pool. In this context, it is also well recognized that as sexually reproducing species evolve, the disparity in size between the two sexes decreases—the male-specific culling mechanisms described above may help to explain this evolutionary phenomenon.

This hormonal strategy for evolving the "second sex" would be expected to have built-in safeguards against males producing excessive amounts of androgens in order to protect the species against such predictably aggressive males. This may also be a biologic strategy for selecting for males with optimal genetic traits for the survival of the species. The experimental model for merging the genetic and endocrine mechanisms may be the B10/B10A congenic strains of mice, which are characterized by spontaneous differences in glucocorticoid-sensitive developmental mechanisms such as cleft palate and lung maturation. Significantly, it is the slower-developing B10A males that exhibit increased androgen activity as adults, as predicted by the genetic–endocrine interactive model. Nature appears to have evolved a comprehensive genetic–endocrine mechanism for the second sex, while at the same time damping this process using the very same genetic–endocrine strategy.

EVIDENCE FOR AN ASSOCIATION BETWEEN STEROID-RESISTANT/ RESPONSIVE PHENOTYPES AND HUMAN LYMPHOCYTE ANTIGEN (HLA) HAPLOTYPES

Observational studies have linked cortisol sensitivity with genetic variation related to the major histocompatability complex (MHC). Ironically, these descriptive studies have now led to a mechanistic reconciliation of the phenotype and genotype, providing a unifying mechanism at the cellular–molecular level for this process. Becker et al. (1976) first reported that different human lymphocyte antigens (HLA, chromosome 6) of the MHC are associated with differential sensitivity to cortisol. Bonner and Slavkin (1975) previously identified two congenic strains of mice, B10 and B10A, which exhibited differential sensitivities to induction of cleft palate by cortisol treatment, providing a genetic link between the endocrine phenotype and the heterochronic development of the palate in these two strains of mice. This was a key observation because these two strains of mice differ genetically only with respect to their H2 haplotype, representing relatively few genes that may be functionally related to postimplantation development within this locus. Studies conducted in Slavkin's laboratory demonstrated that in addition to palatal development, lung development was also heterochronic in these two strains of mice, and that the slower-developing B10A strain was resistant to the acceleratory effect of both exogenous and endogenous cortisol.

Jaskoll and Melnick (2000) have identified the gene product that mediates heterochronic development of the palate and lung in the B10/B10A congenic mouse strains; they found that TGFβ is differentially produced during development in both lung and palate, as it is elaborated longer during gestation in the B10A than in the B10 strain. Since TGFβ blocks epithelial type II cell maturation, the development of these cells is delayed in the B10A strain, resulting in delayed surfactant production. Dominance of the developing lung alveolar acinus by TGFβ results in prolonged lung growth and delayed lung differentiation. The exploitation of TGFβ by both the endocrine and genetic mechanisms of heterochrony is an example of evolutionary convergence; it is particularly striking, considering that in both cases they are terminal differentiation processes, that is, palatal closure (determined by mesenchymal differentiation) and surfactant production (determined by epithelial differentiation), which independently result in heterochrony. The use of the ubiquitous TGFβ-regulated differentiation pathway observed to "time" development in many tissues may reflect different HLA haplotypes. Consistent with this hypothesis, there is clinical evidence that premature infants both with and without RDS differ in their HLA haplotypes; in that study it was observed that the RDS infants were comprised by two specific HLA haplotypes, namely, A3 and B14, which are distinctly different from the haplotypes in the non-RDS group. The relationship between TGFβ, endocrine, and genetic mechanisms of heterochrony during fetal lung development is depicted in Figure 4.3.

It was the phenotypic linkage of human HLA haplotypes to their differential glucocorticoid sensitivities that first led investigators to the B10/B10A phenotype relationship. The observation that the timing of palatal and lung development is the result of TGFβ dynamics unifies the endocrine and genetic heterochronic models mechanistically.

The identification of TGFβ as the common mechanism for both the endocrine and genetic delay of lung development merges these two models into one unified mechanism. This is highly significant because it links the endocrine pathways

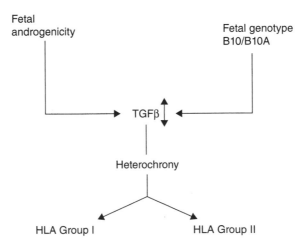

Figure 4.3. Heterochronic development and maturation of specific fetal tissues may arise through differential regulation of TGFβ expression and activity between females and males. Both the fetal androgen state and the fetal genotype may contribute to the differential regulation of TGFβ. In addition, HLA may influence the outcome of the heterochronicity of TGFβ timing of developmental events.

timing lung development to HLA haplotypes associated with functionally relevant candidate genes controlling the process of lung development.

We must address the evolutionary origins of human physiology based on phylogenetic and developmental mechanisms, aided by the human genome. The conventional way of using genomics is through statistical analysis of data, but evolution did not occur by chance. Alternatively, the functional genomic evolutionary approach that we have proposed may fail to directly identify such first principles because we are missing intermediate steps from the molecular "fossil record." But some aspects of those intermediate steps were likely incorporated into other existing functional phenotypes, or into other molecularly related functional homologies, such as those of the lung and kidney, photoreceptors and circadian rhythms, the lens of the eye, and liver enzymes. What this approach does provide is a robust way of formulating refutable hypotheses to determine the ultimate origins and first principles of physiology by providing candidate genes for phenotypes hypothesized to have mediated evolutionary changes in structure and/or function. It also forms the basis for predictive medicine, and even for molecular bioethics, rather than merely showing associations between genes and pathology, which is an unequivocal just-so story. In this new age of genomics, our reach must exceed our grasp.

In this chapter we have shown how the cell–cell communication model of vertebrate evolution predicts the cycle from ontogeny and phylogeny to homeostasis and repair using the lung as a prototype. This cell-biology-oriented model of evolution also predicts the finite nature of the vertebrate lifecycle according to the premise that primitive cells provided an environment in which entropy could be reduced, leading to the formation of multicellular organisms optimized for reproductive success, where the energetic tradeoff is death. In Chapter 5 we will demonstrate how the lung cell–cell communication model predicts the evolution of the lung in response to iterative external and internal selection pressures. Of course, this model is based on the genetic signaling of surviving organisms, and therefore accounts for the gaps in the evolutionary history of vertebrates created by the predictable periodic extinction events caused by transitional organisms that could not adapt under selection pressure.

REFERENCES

Adamson IY, King GM (1984), Sex-related differences in cellular composition and surfactant synthesis of developing fetal rat lungs. *Am. Rev. Respir. Dis.* 129(1):130–134.

Bacallao R, Fine LG (1989), Molecular events in the organization of renal tubular epithelium: From nephrogenesis to regeneration. *Am. J. Physiol.* 257:F913–F924.

Becker B, Shin DH, Palmberg PF, Waltman SR (1976), HLA antigens and corticosteroid response. *Science* 194(4272):1427–1428.

Bonner JJ, Slavkin HC (1975), Cleft palate susceptibility linked to histocompatability-2 (H-2) in the mouse. *Immunogenetics* 2:213–218.

Crespi EJ, Denver RJ (2006), Leptin (ob gene) of the South African clawed frog *Xenopus laevis*. *Proc. Natl. Acad. Sci. U. S. A.* 103(26):10092–10097.

Daniels CB, Orgeig S (2003), Pulmonary surfactant: The key to the evolution of air breathing. *News Physiol. Sci.* 18:151–157.

Dobzhansky T (1973), Nothing in biology makes sense except in the light of evolution. *Am. Biol. Teacher* 35:125–129.

Jablonka E, Lamb MJ (2006), The evolution of information in the major transitions. *J. Theor. Biol.* 239:236–246.

Jaskoll T, Melnick M (2000), The genetics of glucocorticoid-regulated embryonic lung morphogenesis: A first approximation of the epigenetic rules, in Mendelson CR, ed., *Endocrinology of the Lung*. Humana Press, Totawa, NJ.

Kotas RV, Avery ME (1980), The influence of sex on fetal rabbit lung maturation and on the response to glucocorticoid. *Am. Rev. Respir. Dis.* 121(2):377–380.

Nielsen HC (1985), Androgen receptors influence the production of pulmonary surfactant in the testicular feminization mouse fetus. *J. Clin. Invest.* 76(1):177–181.

Polanyi M (1968), Life's irreducible structure. Live mechanisms and information in DNA are boundary conditions with a sequence of boundaries above them. *Science* 160:1308–1312.

Prigogene I (1984), *Order Out of Chaos*. Bantam Books, New York.

Smith JM, Szathmáry E (1995), *The Major Transitions in Evolution*, Oxford University Press, Oxford, UK.

Smolin L (1997), *The Life of the Cosmos*. Oxford University Press, Oxford, UK.

Sweezey NB, Ghibu F, Gagnon S, Schotman E, Hamid Q (1998), Glucocorticoid receptor mRNA and protein in fetal rat lung in vivo: modulation by glucocorticoid and androgen. *Am. J. Physiol.* 275(1 Pt. 1):L103–L109.

Torday JS, Ihida-Stansbury K, Rehan VK (2009), Leptin stimulates *Xenopus* lung development: evolution in a dish. *Evol. Dev.* 11(2):219–224.

West JB, Elliott AR, Guy HJ, Prisk GK (1997), Pulmonary function in space. *JAMA* 277(24):1957–1961.

Wilson EO (1998), *Consilience*. Random House. New York.

5

HOW TO INTEGRATE CELL–MOLECULAR DEVELOPMENT, HOMEOSTASIS, ECOLOGY, AND EVOLUTIONARY BIOLOGY: THE MISSING LINKS

The evolutionary biology literature is replete with heated debates, ranging from what the evolutionary process is, to its very nature as either gradual or episodic, how organisms adapt to their environment, how novelty occurs, and the meaning of epigenesis, to cite just a few of the many examples. Defining epigenesis may bring resolution to these other issues, because it is a way of thinking about how organisms interact with the environment to generate new structures and functions.

In Chapter 4, we made the case for a biologic continuum from development to homeostasis, regeneration, and aging based on the cellular–molecular principles

Evolutionary Biology, Cell–Cell Communication, and Complex Disease, First Edition.
John S. Torday and Virender K. Rehan.
© 2012 Wiley-Blackwell. Published 2012 by John Wiley & Sons, Inc.

developed in Chapters 1–3. Yet, the purpose and intent of this book is to provide a way of thinking about evolutionary biology that is experimentally testable and refutable. In this chapter, we will provide a novel way of functionally integrating ontogeny, phylogeny, and ecology that provides insights into the causal mechanisms and processes underlying evolution.

NEUTRAL THEORY VERSUS INTELLIGENT DESIGN

At the two extremes of evolution lie Kimura's neutral theory and creationism, or intelligent design. In between, there are many speculations as to how evolution has occurred, with little more than natural selection and genetic mutation for mechanisms of macro- and microevolution, respectively. But what does *natural selection* mean? It is a metaphor for *descent with modification*, which lacks a specific mechanism for generating diversity and extinction. Borrowing from the concept of evolutionary developmental biology, contemporary cellular—molecular embryology is based on cell–cell communication as the basis for morphogenesis. Cells produce soluble growth factors, and their cellular signaling partners express the complementary specific receptors that mediate the cell–cell signaling that generates the structural–functional phenotypes. We have intimated that this mechanism functionally integrates development, homeostasis, regeneration, and aging, and as such, it *is* the mechanism of evolution. Viewed at in the forward direction, the interactions of external and internal forces can act on such an integrated regulatory mechanism to cause the evolution of phenotypes. Furthermore, by comprehensively affecting the reproductive capacity of the adults, the development of the progeny, and the global health and well-being of all the individuals within a species, the emergent phenotype is adaptive because it is constrained by internal physiology, and is contingent on the mechanisms that have mediated its evolution.

INTERNAL SELECTION THEORY

In addition to the concepts of natural selection and mutation, there is a third means of evolution that lurks in the corners of the evolutionary literature, namely, internal selection, although that concept is not very popular for two reasons—because it is reminiscent of Lamarckism and because it has been very difficult to demonstrate experimentally, which may be due to the lack of a cellular perspective. We come closest to this phenomenon in the burgeoning discipline of ecological–developmental (eco-devo), but fail to recognize that this may be an aspect of the larger domain of cell–cell communication. This perspective is also embedded in the emerging ideas surrounding cooption, or shifts in the function of a trait during evolution. Such changes may occur in parts of the amino acid sequence not required for contemporary function; changes may occur in gene regulation, expression of a protein in novel tissues, and/or developmental stages; and gene duplication events may occur, followed by divergence of amino acid sequences and/or regulatory DNA sequences of the descendant paralogs. What is lacking is an integration of these molecular communications that have resulted in evolution of phenotypes in a manner that is consistent with, and mediated by, selection pressure. Instead, evolutionists default to natural selection as the mechanism of evolution, thus relying on macroevolutionary processes. This is unfortunate, because there is an important causal interrelationship between macro- and microevolution that is being missed because of the lack of a unifying mechanism. In support of that notion, the schematic in Figure 5.1 showing the confluence of ontogeny, phylogeny, and ecology presented in this chapter provides such a unifying mechanistic perspective, showing causal relationships between all three aspects of the evolutionary strategy. This portal into the core mechanisms of evolution was unavailable to Darwin. It may lead to the in-depth understanding of our evolutionary origins that we have been looking for.

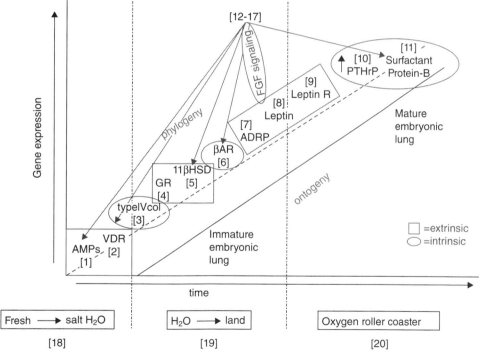

Figure 5.1. Alternating extrinsic and intrinsic selection pressures for the genes of lung phylogeny and ontogeny. The effects of the extrinsic factors (salinity, land nutrients, oxygen) on genes that determine the phylogeny and ontogeny of the mammalian lung alternate sequentially with the intrinsic genetic factors, highlighted by the circles and squares. (See insert for color representation.)

The term *natural selection* infers that nature determines which individuals survive. But what is nature, if not the set of physical laws that biologic systems must ultimately abide by? Like Einstein's famous quip regarding the nature of the physical world, that God does not play dice with the universe, biology derives its properties from the physical world, and is therefore also not a result of chance. We have previously addressed how biology has adapted to its physical environment through communication: unicellular organisms communicating with nature, cell–cell communication to generate physiologic traits, and communication of genetic information from generation to generation during reproduction—as a way of perpetuating the species, inferring that there are both internal and external selection pressures that somehow coordinately determine adaptation.

Now we will invoke the interactions between external and internal selection pressures as a unifying principle for evolution. This is conventionally referred to as *orthogenesis*, which has been unpopular ever since its invocation by Lamarck, who suggested that internal adaptation was a conscious and deliberate mechanism, rather than a passive consequence of cell–cell communication. It is unfortunate that Lamarck chose to characterize evolution in this way, since we recognize that there are patterns of evolution that seem to emanate from internal properties— what Aristotle referred to as *entelechy*, a factor that directs the individual regularities of organisms; what Riedl (1978) referred to as *burden*, the responsibility carried by a feature or decision; what Arthur called *bias* (2004); what L. L. Whyte referred to as *coordinative conditions* ("The coordinative conditions hold the clue to the relation of physical laws to organic processes and to the unity of the organism") (1965). Many have invoked the existence of an internal principle as the basic problem of evolutionary theory—Remane, Ludwig, Hennig, and Hartmann— without providing a mechanism. Yet, we know that evolution is not a conscious process. In this chapter, we will provide evidence for the capacity to generate form and function via the interplay between external and internal selection pressures, mediated by cell–cell interactions.

The representatives of synthetic theory, on the other hand, hold that no causal mechanism has yet been prove. They are also concerned that the search for it would allow for such unproven phenomena as finality and entelechy. Entelechy, a concept taken from Aristotelian metaphysics, is assumed to be a factor that directs the individual regularities of organisms, specifically, their orderliness, harmony, plan, or goal. Entelechy would arise from the preestablished harmony of living organisms, which is outside the realm of science. Instead, they tend to discount the problem, which is unfortunate. In some cases, they even claim that there is no place in the modern synthesis for an internal principle. However, it is important to note that even the authorities on this viewpoint, such as Dobzhansky, Kosswig, and Mayr, acknowledge that the epigenetic system confers a fundamental, but not fully understood, ordering effect. They also ask whether this pattern of mutual gene effects will ever be understood (Kosswig 1959), because of its complexity.

THE COUNTERINTUITIVE NATURE OF PHYSIOLOGY AND SOLUTION TO THE DEAD SEA SCROLLS PUZZLE

There are many aspects of physiology that are *counterintuitive*, suggesting that there is an alternative logic for evolution. As mentioned earlier, the endocrine lung, the pineal as a photoreceptor and determinant of circadian rhythms, the lens of the eye, which is composed of stress-related proteins; the kidney integrating blood volume and erythropoiesis; or why the toll gene, which determines the body axis, should have anything to do with host defense are all unsolved puzzles. We must find counterintuitive ways to decode these phenotypes, like the ingenious archeological investigators who used DNA analysis to reassemble the remaining scraps of parchment from the Dead Sea scrolls.

Deciphering the Dead Sea scrolls began enthusiastically with their discovery in a cave in the Judean desert in 1947. But this process slowed rapidly once the first 15 scrolls were translated, with the remaining parchment shards lying in a pile of an estimated 10,000 postage-stamp-sized pieces derived from about 800 other scrolls. These scraps were too decayed to piece together like a jigsaw puzzle, and the writing on them was limited to a few characters per fragment, so they couldn't be connected through their context, either. The project seemed to be at an impasse. But the archeologists cleverly reasoned that the solution to the problem was not in the writings of the ancient scribes, as logic or intuition would dictate, but in the material that the parchment was made from—after all, the scrolls were made of the skins of individual sheep, and there was enough genetic material in each scrap of parchment for the researchers to determine which pieces fit together by matching their DNA content. A similar counterintuitive, the reverse-engineering approach, must be used to solve the puzzle of evolution, although evolution is far more challenging. The DNA solution to the Dead Sea scrolls problem was arithmetic, whereas the genes involved in the evolutionary process are seemingly incalculable, having been permuted and recombined over thousands and thousands of generations to produce novel forms and functions, requiring much more convoluted conceptualization, and equally esoteric mathematical modeling.

Conventionally, such complex problems in evolution have been circumvented using metaphors such as natural selection, survival of the fittest, genetic assimilation, epistasis, and coordinative conditions. But the time for description and metaphor is over, given that we now have the entire genome for the human being, frog, worm, fish, and bird. We must find the mechanistic connections between phenotypes and genes that have generated "endless forms most beautiful".

Since evolution is emergent and contingent, we must use approaches to its understanding that themselves may seem counterintuitive. By reducing ontogeny, phylogeny, and ecology to the molecular level, and interrelating them mechanistically to generate form and function, we can now understand the principles involved in evolution. This fundamental observation could then be used as a cipher for other evolved physiologic traits.

THE CONTINUUM FROM MICROEVOLUTION TO MACROEVOLUTION

By reducing the phenotypic algorithms to their smallest functional components, and then interrelating them through common mechanisms of morphogenesis, we will be able to drill down to the underlying principles of evolution. Similarly, we could disentangle the evolutionary principles behind the puzzles of lens proteins as crystalins, antifreeze proteins, cytoskeletal proteins, butterfly eyespots, Hox genes, and tetrapod limbs. In contemporary biology and medicine, we are faced with the challenge of merging phenotypic patterns with the genomic and molecular biologic data that undoubtedly generated them. These interrelationships are most apparent during development, because the genetic changes that determine morphogenetic phenotypes are causal. We have suggested that developmental–homeostatic interrelationships could be exploited to deconvolute the evolution of the lung as a model for the molecular evolution of the physiologic principles in described Chapters 2–4. The premise that all of biology is a continuum offers the opportunity to exploit that observation in order to determine where it began, and how it was perpetuated through evolution. In the present chapter, we have expanded on this concept by integrating the ontogenetic and phylogenetic sequence of genetic mechanisms, regressing them against the temporal sequence of specific geoenvironmental milestones in vertebrate evolution: increased ocean salinity, terrestrial nutrients, and atmospheric oxygen. The potential causal relationships delineated in this model have been suggested by their relevance to known developmental mechanisms. The inference in the schematic in Figure 5.1 is that the ontogeny and phylogeny of the lung (lying along the y axis) have been determined by the impact of extrinsic environmental factors (lying along the x axis) on intrinsic cellular-molecular mechanisms, interacting to generate lung structure and function through positive selection pressure—a process that relates all the way back to the origins of life as micelles forming in response to environmental factors (see Chapter 1). All of the molecular steps in lung development employed in the model have been shown to be causally related experimentally, and in some cases the phenomena can be mechanistically explained based on the basis of the underlying nature of the mediators. For example, salinity inhibits innate host defense in fish, but stimulates it through increased vitamin D metabolism, acting as a balancing selection pressure to facilitate evolutionary adaptation. Other examples are the selection pressure for type IV collagen synthesis; the effect of glycerrhetinic acid, the product of rancidification of land vegetation, on the specialization of the mineralocorticoid and glucocorticoid receptors—pentacyclic triterpenoids such as glycerrhetinic acid inhibiting 11β-hydroxysteroid dehydrogenase (11βHSD) 1 (11βHSD1), causing increased blood pressure, resulting in balancing selection for both the glucocorticoid receptor and 11βHSD1, resulting in local activation of cortisol within cells and tissues; or the impact of fluctuating oxygen tension in the environment over the past ~500 million years on the differentiation of the lipofibroblast to protect the alveolar wall against oxidant injury. It would be expected that if this is positive selection, then the emerging adaptations would also positively interact with the earlier adaptations; in other words, glucocorticoids have a positive effect on the vitamin D receptor. Oxygen positively affects glucocorticoid signaling and the vitamin D receptor—perhaps this is seen best as the effect of oxygen on the emergence of the lipofibroblast, producing leptin, which then pleiotropically affects antimicrobial peptides (AMPs), the vitamin D receptor (VDR), type IV collagen, and the surfactant, thereby acting to facilitate the expansion of the blood–gas barrier.

Salinity and the Vitamin D Receptor

The speculation that external selection pressure has primarily been on the lung innate host defense system is reinforced by the first example of an external selection pressure affecting molecular lung development and evolution, as follows. The increase in salt content of the oceans (see Fig. 5.1, step 18), possibly a concomitant result of the desiccation of water environments thought to have driven vertebrates onto land (the Romer hypothesis), is depicted in Figure 5.1 on the x axis (abscissa) as a geochemical selection pressure for lung evolution. The hypothesized mechanism of action of salinity resulted from its inhibition of AMP bioactivity (Fig. 5.1, step 1), counterbalanced by increased vitamin D signaling (Fig. 5.1, step 2), which independently stimulates tissue-specific innate host defense, since fish exposed to elevated salinity show coordinated, adaptive changes in both vitamin D hydroxylation (Sundh et al. 2007) and the corresponding vitamin D receptor (Larsson et al. 2003). This selection pressure for host defense may, in turn, have referred back to the anatomy of the physostomous fish swim bladder, which is connected to the gut tube by the pneumatic duct, allowing the fish to inflate its swim bladder by rising to the surface and taking in air shortly after hatching. This adaptation to surface for air may have been the antecedent of the transition by this class of fish from water to land. It may also have created the positive selection pressure for host defense, since bacteria could have entered the swim bladder via the pneumatic duct. In support of this concept, leptin produced in the alveolus, and lipopolysaccharide, the major component of the outer membrane of Gram-negative bacteria, have similar effects on type II cell differentiation, suggesting that the constitutive evolutionary selection pressure may have been for host defense, and that leptin evolved into the intrinsic mechanism for regulated host defense and barrier function. Such preadaptive balancing selection for the regulation of innate host defense in combination with vitamin D regulation of blood pressure may have resulted in the genetic shift in the P450 cytochrome CYP3A from a salt-sensitive to a salt-resistant form, facilitating the migration of humans away from the equator (Thompson et al. 2004). In fact, it has been found that the salt-resistant form of CYP3A is among genes that have recently rapidly evolved in humans. This may have been a consequence of domestic salt use by early humans, inadvertently causing both increased autoimmune diseases due to inhibition of antimicrobial peptides, and elevated blood pressure due to the maladaptive response to salt. Such observations are supported by experimentally deleting both the vitamin D receptor and 1α-hydroxylase, which activates vitamin D in mice, resulting in elevated blood pressure; an effect that was reversed by feeding the animals vitamin D. In other such studies, the immune response is similarly affected.

The proximate result of such selection pressure for vitamin D metabolism may have been the northerly exodus of progressively lighter-skinned people out of Africa. The distal result may have been the acquisition of autoimmune diseases and the propensity to elevated blood pressure; for each 10° north or south of the equator, blood pressure increases by 2.5 mmHg (millimeters of mercury) and hypertension prevalence increases by 2.5%. As an aside, a Boolean search for vitamin D and blood pressure yielded 888 hits and for immunity, yielded 1186 hits.

This scenario is consistent with the racial and ethnic incidences of hypertension, which may be linked to the evolutionary significance of other cytochrome P450 genes that cause disease susceptibility or resistance. Among these, CYP3A may represent the most important P450 gene family, because of the wide substrate specificity of these enzymes, induction by many commonly used drugs, and its high level of expression in the liver and gastrointestinal tract. Such a developmental–evolutionary approach to population genetic data reflects the robust power of the cell-centric understanding of the first principles of physiology advocated in this book.

Type IV Collagen and Increased Surface Area of the Lung

The next major environmental selection pressure occurred when vertebrates transitioned from water to land (Fig. 5.1, step 19) approximately 300 million years ago, causing selection pressure for type IV collagen (Fig. 5.1, step 3), which acts to physically stent the walls of the lung airsacs, or alveoli. We know from studies of Goodpasture's syndrome (MacDonald et al. 2006) that the 3α isoform of type IV collagen evolved sometime between the emergence of fish and amphibians through selection pressure for specific amino acid substitutions that rendered it more hydrophobic and negatively charged, physically preventing the exudation of water and proteins from the microcirculation into the alveolar space.

Goodpasture's syndrome is an autoimmune disease caused by simultaneous kidney and lung failure, caused by pathogenic circulating autoantibodies targeted to a set of discontinuous epitope sequences within the noncollagenous domain 1 (NC1) of the α3 chain of type IV collagen [α3(IV)NC1], referred to as the *Goodpasture autoantigen*. Basement membrane extracted NC1 domain preparations from *Caenorhabditis elegans*, *Drosophila melanogaster*, and *Danio rerio* do not bind Goodpasture autoantibodies, while frog, chicken, mouse, and human α3(IV) NC1 domains bind autoantibodies. The α3(IV) chain is not present in worms (*C. elegans*) or flies (*Drosophila melanogaster*), and is first detected in fish (*Danio rerio*). Interestingly, native *D. rerio* α3(IV)NC1 does not bind Goodpasture autoantibodies. In contrast to the recombinant human α3(IV)NC1 domain, there is complete absence of autoantibody binding to recombinant *D. rerio* α3(IV)NC1. Three-dimensional molecular modeling of the human NC1 domain suggests that evolutionary alteration of electrostatic charge and polarity due to the emergence of critical serine, aspartic acid, and lysine amino acid residues, accompanied by the loss of asparagine and glutamine, contributes to the emergence of the two major Goodpasture epitopes on the human α3(IV)NC1 domain, as it evolved from *D. rerio* over 450 million years. The evolved α3(IV)NC1 domain forms a natural physicochemical barrier against the exudation of serum and proteins from the circulation into the alveoli or glomeruli, due to its hydrophobic and electrostatic properties, respectively, which were more than likely the molecular selection pressure for the evolution of this protein, given the oncotic and physical pressures on the evolving barriers of both the lung and kidney.

Differentiation of the Glucocorticoid Receptor from the Mineralocorticoid Receptor

The other more recently discovered evolutionary effect of the transition from water to land on vertebrate physiology was the specialization of the steroid hormone receptor into the mineralo- and glucocorticoid receptors (Fig. 5.1, step 4). The primary selection pressure may have been due to the rise in ocean salinity, which was accommodated structurally by type IV collagen, combined with sodium regulation by the steroid hormone receptor mineralocorticoid activity. But the subsequent counterbalancing selection pressure may have been caused by the ingestion of pentacyclic triterpenoids by land vertebrates. Such compounds are produced by the rancidification of carbohydrates, a process that is unique to land vegetation. Pentacyclic triterpenoids inhibit glucocorticoid inactivation by 11βHSD2 (Fig. 5.1, step 5), causing glucocorticoid stimulation of mineralocorticoid receptors and elevated blood pressure. This would have created positive selection for both 11βHSD1, which inactivates glucocorticoids, counterbalancing the blood pressure–elevating effect of the mineralocorticoids (Fig. 5.1, step 5), and the specialization of the mineralo- and glucocorticoid receptors through two amino acid substitutions in the steroid binding site of the glucocorticoid receptor in tetrapods.

The Role of βARs in Increased Lung Surface Area

The combined positive selection pressure for 11βHSD1 by salt upregulation of the VDR (see Fig. 5.1, steps 2 and 17) and by the pentacyclic triterpenoid inhibition of 11βHSD2 (see Fig. 5.1, steps 5 and 18) may have led to the subsequent positive selection pressure for lung βARs (Fig. 5.1, step 6) independently regulating pulmonary and systemic blood pressure. Lipofibroblasts subsequently appeared in response to rising atmospheric oxygen tension (Fig. 5.1, step 19); lipofibroblast

11βHSD1 may have been constitutively upregulated by vitamin D_3 produced by lung epithelial cells in response to salt, leading to constitutive lipofibroblast leptin expression (Fig. 5.1, step 8), which would have acted to further stimulate AMPs (Fig. 5.1, step 1), type IV collagen (Fig. 5.1, step 3), and surfactant production (Fig. 5.1, step 10), synergistically increasing barrier function at multiple levels as a stable platform for further increases in lung surface area.

Atmospheric Oxygen, Adipocytes, Surfactant, Body Temperature, and Lung Evolution

Over the course of vertebrate evolution during the Phanerozoic period (the last 500 million years), the amount of oxygen in the atmosphere has increased from 12% to 21%. However, it did not increase linearly, but instead it increased and decreased several times, reaching concentrations as high as 35%, falling to as low as 15% over this time period. The increased oxygen tension may have caused the differentiation of muscle cells into lipofibroblasts in the lung, as the first anatomic site where the increased atmospheric oxygen would have affected selection pressure for evolutionary change, since muscle stem cells will spontaneously differentiate into adipocytes in 21% oxygen (room air), but not in 6% oxygen, suggesting that as the atmospheric oxygen tension increased over evolutionary time, lipofibroblasts could have formed spontaneously. This would have been particularly relevant to the muscle cells surrounding the lung airsacs, which are exposed to the air environment. Consistent with this hypothesis, we have previously shown that the lipids stored in alveolar lipofibroblasts protect the lung against oxygen injury. In turn, the leptin secreted by the lipofibroblasts binds to its receptor on the alveolar epithelial cells lining the alveoli, stimulating surfactant synthesis. The increased production of surfactant would have reduced the alveolar surface tension, resulting in a more deformable gas exchange surface on which selection pressure could ultimately select for the stretch-regulated PTHrP coregulation of surfactant and microvascular perfusion. This mechanism could have ultimately given rise to the mammalian lung alveolus, with maximal gas exchange resulting from coordinate stretch-regulated surfactant production and alveolar capillary perfusion, thinner alveolar walls due to PTHrP's apoptotic or programmed cell death effect on fibroblasts, and a blood–gas barrier reinforced by type IV collagen. This last feature may have contributed generally to the molecular bauplan for the peripheral microvasculature of evolving vertebrates.

As a result of the increased oxygen exposure, the lipofibroblasts of the lung alveoli were able to accumulate fat, mediated by adipocyte differentiation–related protein (ADRP) (Fig. 5.1, step 7), which protects the lung against oxygen toxicity. These lipofibroblasts produce leptin (Fig. 5.1, step 8), which stimulates surfactant production by their neighboring epithelial cells (Fig. 5.1, step 11), making the alveolar sacs more compliant, or "stretchy," permitting more efficient oxygen exchange between the air and the systemic circulation, further increasing tissue oxygenation for metabolic drive. One consequence of this may have been the induction of fat cells in the peripheral circulation, which led to endothermy, or warm-bloodedness. The increase in body temperature would have further increased lung oxygenation because the surfactant is 300% more active at 37°C than at atmospheric temperature (body temperature for cold-blooded organisms). These major physiologic changes in response to the increase in atmospheric oxygen would have been severely challenged by the subsequent hypoxic conditions resulting from the relative decrease in oxygen tension, perhaps being accommodated in the survivors of such mass extinctions by increased βAR production, since hypoxia is the most potent physiologic stress stimulus for adrenal epinephrine production, facilitating the coevolution of pulmonary and adrenal systems. We speculate that this may also have led to selection pressure for the on-demand alveolar homeostasis in reciprocating breathers, that is, stretch regulation of barrier function (surfactant production, AMPs and type IV collagen) in combination with alveolar capillary perfusion through the coordinate upregulation of PTHrP and leptin. The preadaptations to saltwater (Fig. 5.1, step 18), and to the water-to-land transition (Fig. 5.1, step 19) provided the lung with the genetic and epigenetic means for the organism to survive and adapt to the oxygen fluctuations (Fig. 5.1, step 20) that have occurred over the last 500 million years.

Predation, PTHrP, Leptin, and Stretch-Regulated Alveolar Homeostasis

As a result of the evolution of increased oxygenation, the transitional tetrapods (amphibians and reptiles) would have been more metabolically active, putting additional pressure for selection on the lung to evolve even more efficient gas exchange. This was achieved by development of the stretch-regulated surfactant system, in which the stretching of the alveolar wall increased PTHrP production by the epithelial cells (Fig. 5.1, step 10), stimulating leptin production (Fig. 5.1, step 8) by the lipofibroblasts, causing more surfactant production and further increased stretching of the alveoli, creating further selection pressure for stretch regulation of the leptin receptor (Fig. 5.1, step 9). This PTHrP–leptin stretch-regulated mechanism was reinforced by PTHrP stimulation of bloodflow through the alveolus, since PTHrP is also a potent vasodilator, acting to further facilitate gas exchange in synchrony with increased surfactant facilitation of alveolar wall stretching, driving positive selection pressure. Leptin production further reinforced all of these evolutionary steps (Fig. 5.1, steps 12–17) by coordinately stimulating the formation of antimicrobial peptides (Fig. 5.1, step 1), protecting the increased surface area of the lung against infectious agents, and coordinately increasing type IV collagen (Fig. 5.1, step 3) and surfactant production (Fig. 5.1, step 10), which would have prevented rupture of the alveolar blood–gas barrier under physiologic stress as vertebrates evolved from cold- to warm-blooded animals, and from more preylike to more predatorlike animals (fight or flight).

The reinforcing effects of leptin for alveolar homeostasis may represent epis-tasis, since each of these terms for physiologic traits—barrier function, host defense, and alveolar compliance—refers to the deepest homologies expressed all the way back to unicellular organisms. The earliest mechanism by which eukary-otes evolved from prokaryotes was through the formation of a nuclear envelope as a barrier to protect its DNA, or host defense. And we have previously shown that the stretch signaling by PTHrP is an adaptation to gravitational force, which is undoubtedly among the most ancient effectors of biologic adaptation, because they were present in the environment even before oceans, salinity, or oxygen. In fact, we have shown that this is a property of single cells experimentally. When lung and bone cells are placed in a microgravitational environment, the PTHrP messenger RNA levels decrease; when these cells are put back into unit gravity, the PTHrP messenger RNA level returns back to normal. To demonstrate the biologic significance of this gravitational effect on PTHrP expression, we further showed that PTHrP messenger RNA is decreased in the bones of rats that had been in deep space for 2 weeks. This finding was significant because microgravity causes osteoporosis in astronauts, potentially as a result of decreased PTHrP in bone. This effect of gravity on PTHrP may also be the underlying cause for the documented lack of normal alveolar gas distribution in the lungs of astronauts, given that PTHrP coordinates lung alveolar ventilation/perfusion matching.

It is noteworthy that each of the genetic steps in lung evolution depicted in Figure 5.1 (steps 1–11) alternately clusters as either an external environmental (extrinsic, red squares) or an internal (intrinsic, red circle) factor, which integrates environmental and genetic factors to evolve the vertebrate lung. Note that both the direction of change and the sequence in which these genes appear are the way they appear phylogenetically from fish to frogs, to reptiles and humans, and they become mechanistically integrated and regulated through cell–cell interactions during lung development and evolution.

It has been suggested that such interrelationships may just be the result of the human mind seeking logic in nature as an anthropomorphism. That is not true for the current analysis since both the sequence of genetic changes during lung devel-opment and the geologic environmental changes are well-established facts. Fur-thermore, the causal relationships that we have outlined have all been determined experimentally.

A central tenet of this model is that cell–cell interactions underlie the funda-mental mechanisms of vertebrate evolution. On the basis of that precept, it is conceivable that the inhibition of such mechanisms, particularly using specific agents that mimic the primordial factors that generated the evolutionary pheno-types, could reverse these putative evolutionary traits. It is reaffirming, therefore, that such diverse agents as oxygen, pressure (as a proxy for gravity), infection, and nicotine—the chemical mediator of the effects of cigarette smoke—all act to simplify the lung in a reverse evolutionary direction. These observations are perhaps more significant because they provide novel ways of thinking about the effective treatment of chronic lung disease.

This working model of lung evolution was prompted by the realization that at least three known effectors of lung development mechanistically connect phylogeny to major environmental factors: evolution of the vitamin D hydroxylation mechanism and salinity, evolution of the glucocorticoid receptor and the water-to-land transition, and advent of the lipofibroblast and oxygen in the atmosphere.

By *reverse-engineering* lung surfactant, we have discovered the interactions between the environment and the organism that have brought about the evolution of the lung. E. B. Lewis' discovery that genes that controlled the identity of the abdominal segments of the fly were clustered on chromosomes, and in the same sequence as they were expressed along the anteroposterior body axis. He called this phenomenon *collinearity*. Is the interrelationship between the geologic environmental changes and the gene regulatory network for lung development also an example of collinearity? Or does the latter provide the link to the former? Since the lung also shows an Anteroposterior (A/P) pattern, are there Hox genes that determine the lung A/P pattern, and do they demonstrate collinearity?

The inference of this model of lung evolution driven by extrinsic and intrinsic factors is that the external selection pressures *caused* the internal physiologic adaptations in response to these interactions. The genetically regulated mechanisms originate as housekeeping genes, and appear to evolve into regulated paracrine mechanisms in response to deep homologic selection pressures linked together through the properties of cholesterol. Also, by implication, the cellular-molecular mechanisms facilitated the evolution of the lung phenotype in those organisms that were able to molecularly adapt, whereas those organisms that could not became extinct, and therefore are not represented in this analysis. By eliminating the timescales over which these genes have evolved (development, homeostasis/regeneration, reproduction, aging), we have been able to delineate their putative roles in the evolution of the blood–gas barrier. This application of parsimony for deriving the interrelationships between genes and phenotypes as evolutionary processes is applicable to all tissues and organs.

This model of lung evolution not only merges the genetic mechanisms of ontogeny and phylogeny with ecology, but also lends itself to a mechanistic way of merging such seemingly disparate processes as gradualism and punctuated equilibrium as a continuum. Under what might be regarded of as normal variation in environmental selection pressure, there may have been gradual changes in genetic expression of housekeeping genes in response to natural selection, not unlike those described for eco-devo mechanisms (Gilbert and Epel 2009). However, under conditions of severe, inescapable global selection pressures that would have threatened the entire species with extinction, like those depicted in the schematic in Figure 5.1, major adaptive strategies would have forced the emergence of paracrine signaling mechanisms derived from earlier, gradualist genetic amplifications—evolve or become extinct!

Experimental evidence for the above comes from Thornton and his colleagues (Bridgham et al. 2006), who have shown that the glucocorticoid receptor evolved from the mineralocorticoid receptor just as vertebrates emerged from water onto land, seemingly as luck would have it. However, this is a just-so story that does not address what the selection pressure on the hormone receptor and 11βHSD mechanisms was, and how it facilitated physiologic adaptation.

The cascade described in Figure 5.1 begins with balancing selection for host defense, inhibited by salinity, counterbalanced by increased vitamin D metabolism to stimulate antimicrobial activity. Since vitamin D is synthesized from cholesterol, this mechanism further references the earlier properties of cholesterol for plasma membrane fluidity, combined with its effect on receptor signaling, and its role as a primitive lung surfactant. As a result of these cholesterol-derived properties, the metabolic activation of vitamin D would have facilitated the positive selection for other housekeeping genes such as 11βHSD, the leptin receptor, PTHrP and lung surfactant. This would have provided the preadaptations (or exaptations) that would have facilitated the subsequent effects, starting with adaptation to land, and then for adaptation to the fluctuations in atmospheric oxygen, generating the lipofibroblast, the production of leptin, stimulating lung surfactant production, increased distensibility, positive selection for stretch regulation of PTHrP, culminating in stretch-integrated surfactant production, and ventilation/ perfusion matching—*all of which are emergent and contingent*, which now provides a mechanistic basis for this well-recognized description of the evolutionary process for the first time.

The power of this model is that it is empirically testable and refutable, in accord with Popper's (2002) criteria for a rigorous scientific approach. More importantly, it provides a novel way of thinking about positive selection pressure, given the common pathway from the fish swim bladder to the evolution of the mammalian lung, predicting that the genes in this paracrine pathway will be highly polymorphic, which they all are, providing plasticity for physiologic evolution, on one hand and a basis for understanding the genetic basis for lung diseases on the other hand. The latter may be of value in devising rational drug treatments, and will be discussed in Chapter 10.

Evolutionary biology is a paradoxical reconciliation of the need for both constancy and novelty in homeostatically regulated mechanisms. Historically, Severtsov had coined the term *aromorphosis* as a descriptor for this blackbox mechanism; E. B. Cope had generated the neologism *physiogenesis* to describe the physicochemical changes caused by the environment, and the individual's ability to respond effectively to those changes, a complex process that he referred to as *accommodation*. Jantsch (1980) approaches a mechanistic explanation by stating that the system is a self-organizing, autonomic, autocatalytic process that is self-referential with respect to its own evolution. This is the result of self-generating disequilibrium, as Prigogene had described for inorganic, dynamic systems in *Order Out of Chaos*, in which nonequilibrium can be the source of order, or organization, becoming the basis for nonlinear thermodynamics of irreversible processes that generate *spontaneous structuration*. Jantsch then allows for mechanisms of *coevolution* and *ecological niche complexification*, emphasizing that the behavior of the organism in relation to these factors allows it to not only "determine within relatively wide boundaries to which natural selection it subjects itself" but also participate in further evolutionary progress, by responding to new stresses that arise from its new activities. This sounds like a functional definition of evolvability (Kirschner and Gerhart 1998), and for the first, and to our knowledge, only time suggests the possibility of interactions between the organism and the environment that drive evolution.

Many conceptual scientific breakthroughs to complex problems have resulted from observations of simple relationships, starting with Archimedes' Eureka! moment, when he realized that the displacement of water in his bathtub explained the principle of buoyancy. One example of such an ah-hah moment that many of us had as kids—the recognition that continental drift was due to plate tectonics—came from looking at a globe and realizing that South America and Africa may have been one continent at one time. In his work *Thesaurus Geographicus*, Abraham Ortelius suggested that the Americas were "torn away from Europe and Africa . . . by earthquakes and floods . . . The vestiges of the rupture reveal themselves, if someone brings forward a map of the world and considers carefully the coasts of the three [continents] " (Romm 1994). Or Mandelbrot's discovery of fractals, based on the patterns of dog-eared pages in a text he was studying, not unlike Kary Mullis' discovery of the principle for the polymerase chain reaction while driving up California Highway 101 to Mendocino one foggy night (Yoffe 1994).

By contrast, Darwin spent 5 years sailing on *HMS Beagle* making countless observations of flora, fauna, and geologic formations that he used as the basis for *The Origin of Species*. We continue to puzzle over the mechanisms of natural selection and survival of the fittest. Neither Darwin nor Wallace, nor anyone else to date, has been able to show the specific cause–effect, mechanistic relationships between environmental selection pressures and adaptations. In defense of Darwin and Wallace, they did not have knowledge of genetics during their lifetimes. Even the inventive Riedl, who described in great detail the nature of *Order in Living Organisms*, did not see the connection between the physical and physiologic domains.

The schematic depicted in Figure 5.1 mechanistically integrates environmental selection pressure and cellular-molecular processes for the first time. It displays the well-known sequence of molecular events underlying mammalian lung development on the y axis, and major ecological events in Earth's history over this same period of time on the x axis in the temporal sequence that they occurred. Interestingly, specific environmental changes, beginning with increased ocean salinity (see Fig. 5.1, step 18), followed by the presence of land-specific substances in the diet (see Fig. 5.1, step 19), and the Phanerozoic fluctuations in atmospheric oxygen (see Fig. 5.1, step 20), undoubtedly impacted on the process of lung evolution. By examining these evolutionary milestones systematically, we have discovered that there is an alternating, iterative pattern of external and internal environmental factors (using the term *internal environment* in the same way that Claude Bernard (1974) did, as the *milieu interieur*) that have determined the structural and functional evolution of the lung.

Evolution occurs as a result of genetic adaption to changes in the environment under selection pressure. Conventionally, on the basis of natural selection, macro- and microevolutionary adaptations are viewed as being independent processes, yet the patterns of diversity would suggest that there is an integrated continuum from the proximate to the ultimate mechanisms of evolution. The schematic in Figure 5.1 was generated by regressing the cellular–molecular paracrine mechanisms of vertebrate lung development against major environmental changes that occurred during these periods of vertebrate lung evolution. Using a case study approach, we have demonstrated how environmental factors such as the increase in salinity of the oceans, the presence of pentacyclic triterpenoids resulting from the rancidification of land vegetation, and oxygen fluctuations in the atmosphere have affected cell–cell signaling in a sequence consistent with the ontogeny, phylogeny, and evolution of the lung. The net result of such selection pressure over the evolutionary history of the lung has been the progressive decreases in alveolar septal thickness and alveolar diameter, mediated by *cis* regulatory mechanisms, namely, Wnt/β-catenin and PTHrP/PTHrP receptor paracrine signaling, as depicted in Figure 5.2. Moreover, this selection pressure did not induce new gene expression, but instead facilitated the *cis* regulation of housekeeping genes. In all cases, the causal nature of these changes is documented by experi-

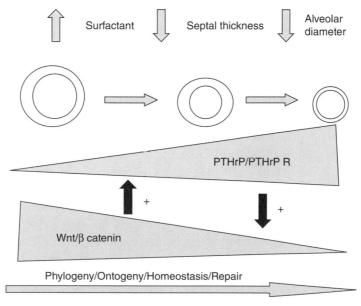

Figure 5.2. Progressive evolutionary decreases in alveolar septal thickness and diameter. Over the evolutionary history of the lung alveolus, its phylogeny and ontogeny have been shaped by selection pressure for increased surfactant production, mediated by Wnt/β-catenin and PTHrP/ PTHrP receptor paracrine signaling. (See insert for color representation.)

mental evidence for these structural and functional traits. Moreover, there is evidence that the more distal changes complement and reinforce the more proximal changes, suggesting that the origins of the selection pressure for the lung may have derived from deep homologies. The overt selection pressure appears to be for the increased efficiency of the surfactant system, allowing for the progressive decrease in alveolar diameter, resulting in the increased area of the gas exchange surface area/blood volume ratio. Yet we have made the paradoxical observation that leptin stimulates this process in the frog tadpole lung in the same way that it does in the mammalian lung (discussed further in Chapter 7), increasing the surface area of the lung in association with stimulation of the surfactant system. However, the frog lung does not require the same gas exchange efficiency as the mammalian lung since the gas exchange surface is composed of faveoli, which are so large in diameter as to render the reduction in surface tension at the faveolar surface unnecessary. These data are more likely the consequence of primary selection pressure for increased expression of innate host defense mechanisms, such as the antimicrobial surfactant proteins A and D. This insight may also explain why bacterial infection seemingly paradoxically stimulates lung development, another example of a counterintuitive phenomenon requiring a big picture evolutionary perspective.

cis REGULATION AND ADAPTIVE EVOLUTION

On the basis of our findings with regard to the intrinsic–extrinsic interactions between the individual and its surroundings, and how that has caused genetic changes consistent with lung development and evolution, we have configured the following integration from micro-to macroevolution (Fig. 5.3). Small ecological

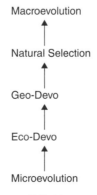

Figure 5.3. Ur scenario.

adaptations affect the evolutionary strategies of the species at multiple levels (physiologic adaptation, individual breeding success, developmental adaptations, and then start from the top again as an iterative process of genetic assimilation). Such well-recognized ecological–developmental (eco-devo) mechanisms (Gilbert and Epel 2009) may provide the selection advantage that occurs when large geologic changes occur, as suggested by our data. Such large-scale selection pressures amount to extinction/survival episodes at the macroevolutionary level. Therefore, there is a demonstrable continuum from the micro- to the macroevolutionary scales.

The model of evolution as a continuum from ontogeny and phylogeny to homeostasis, regeneration and repair, and aging described in this chapter is based on well-established cellular communication mechanisms. As such, these intercellular communications are amenable to changes due to environmental factors at all phases of the lifecycle. Bearing this in mind, we note that it is distinctly possible that environmental change would affect organisms throughout their life history in order to conform with the environment. We know of many examples of adaptation to environmental factors. Such environmental adaptations not only select for those organisms that are plastic but also foster plasticity through selection. Over evolutionary time, this mechanism may have selected for being the most amenable to evolution of the species through epigenetic adaptation, integrated into the biologic composition of the species, and reinforced by the reproductive strategy. If a major geologic change occurred, those organisms that had evolved would have been best suited to survive and thrive in the aftermath. The schematic depicted in Figure 5.1 provides evidence for the existence of such a process, alternating between internal and external selection pressures, facilitated by the biologic data operating system.

By reducing the evolution of the lung to the molecular mechanisms of development, phylogeny, and geologic effectors, we have been able to recapitulate the underlying mechanisms. This approach circumvents the criticism that one cannot discern such mechanisms by looking at the development of extant organisms (Wagner 2001). By testing the hypothesized effect of the environment on cell–cell signaling for ontogeny and phylogeny, we show causality for the first time, unencumbered by this argument.

EVOLUTION OF *cis* REGULATORY MECHANISMS

Elsewhere in this book we have pointed out that one of the hallmarks of the evolution of complex physiology is the transition from housekeeping genes to regulatory genes through cell–cell communication, as depicted in Figure 5.4. The epithelial lining cells of the fish swim bladder express surfactant protein A and secrete cholesterol as unregulated housekeeping functions. In contrast to this, the amphibian lung epithelium expresses surfactant protein B, and surfactant phos-

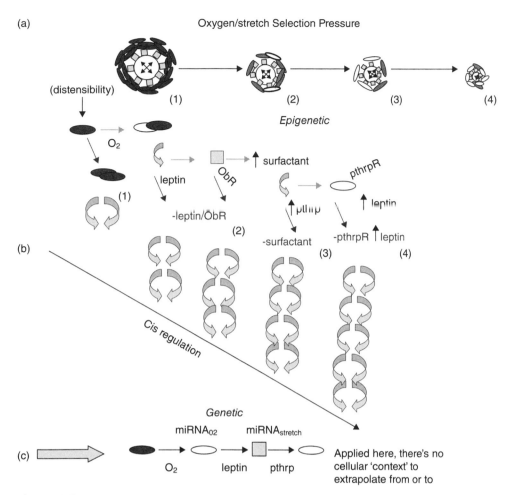

Figure 5.4. Evolutionary–developmental origins of the lung. The top portion **(a)** depicts the paracrine cellular mechanism of alveolar development, phylogeny, and evolution. The middle portion **(b)** depicts the cell–cell signaling mechanisms that have facilitated alveolar evolution, mediated by soluble factors such as leptin and PTHrP, and the *cis* regulatory elements that had to evolve. The cascade shows how oxygen and stretch may have driven a series of cellular interactions for lung evolution, from the advent of the lipofibroblast to leptin production, causing increased surfactant and increased distensibility, placing positive selection pressure on stretch regulation of PTHrP. This process ultimately allowed for the progressive decrease in alveolar diameter, increasing the the surface area/volume ratio and increased gas exchange. Portion **(c)** shows a traditionally descriptive genetic pathway for the same processes shown in (a) and (b). (See insert for color representation.)

pholipids are regulated by leptin (discussed in Chapter 7) produced by lipofibroblasts, which are the key to understanding mammalian lung evolution—the cellular paracrine interactions that mediated lung evolution (depicted in Fig. 5.4a) are underpinned by the evolution of cell–cell signaling through leptin and PTHrP, soluble factors that bind to their cell surface receptors, triggering the expression of *cis* regulatory mechanisms (Fig. 5.4b). This cell–cell communication view of evolution is contrasted with the conventional genetic perspective, which is depicted in Figure 5.4c.

In this chapter, we have reduced the process of evolution to adaptations mediated by *cis* regulatory mechanisms. The role of *cis* regulation in evolution has been described by others (Prud'homme et al. 2007; Wagner and Pyle 2007), but has not been conceived of as an integrated process that predicts changes in structure and function that would account for evolutionary history. In Chapter 6, we will extrapolate from the lung cellular-molecular evolution mechanism to the evolution of other organs, forming the basis for integrated physiology from its origins.

REFERENCES

Arthur W (2004), *Biased Embryos and Evolution*. Cambridge University Press, Cambridge, UK.

Bernard C (1974), *Lectures on the Phenomena Common to Animals and Plants*. Trans. Hoff HE, Guillemin R, Guillemin L. Charles C. Thomas, Springfield, IL.

Bridgham JT, Carroll SM, Thornton JW (2006), Evolution of hormone-receptor complexity by molecular exploitation. *Science* 312(5770):97–101.

Gilbert SF, Epel D (2009), *Ecological Developmental Biology: Integrating Epigenetics, Medicine, and Evolution*. Sinauer Associates, Inc., Sunderland, MA.

Jantsch E (1980), *The Self-Organizing Universe*. Pergamon Press, New York.

Kirschner M, Gerhart J (1998), Evolvability. *Proc Natl Acad Sci USA* 95(15):8420–8427.

Kosswig C (1959), Phylogenetische trends genetisch betrachtet. *Zool. Anz.* 162:208–221.

Larsson D, Nemere I, Aksnes L, Sundell K (2003), Environmental salinity regulates receptor expression, cellular effects, and circulating levels of two antagonizing hormones, 1,25-dihydroxyvitamin D3 and 24,25-dihydroxyvitamin D3, in rainbow trout. *Endocrinology.* 144(2):559–566.

MacDonald BA, Sund M, Grant MA, Pfaff KL, Holthaus K, Zon LI, Kalluri R (2006), Zebrafish to humans: Evolution of the alpha3-chain of type IV collagen and emergence of the autoimmune epitopes associated with Goodpasture syndrome. *Blood* 107(5):1908–915.

Popper KR (2002), *The Logic of Scientific Discovery*. Routledge, London.

Prud'homme B, Gompel N, Carroll SB (2007), Emerging principles of regulatory evolution. *Proc. Natl. Acad. Sci. USA* 104(Suppl. 1):8605–8612.

Riedl R (1978), *Order in Living Organisms*. John Wiley & Sons, Chichester.

Romm J (1994), A new forerunner for continental drift. *Nature* 367(6462):407–408.

Sundh H, Larsson D, Sundell K (2007), Environmental salinity regulates the in vitro production of[3H]-1,25-dihydroxyvitamin D3 and [3H]-24,25 dihydroxyvitamin D3 in rainbow trout. *Gen. Comp. Endocrinol.* 152(2–3):252–258.

Thompson EE, Kuttab-Boulos H, Witonsky D, Yang L, Roe BA, Di Rienzo A (2004), CYP3A variation and the evolution of salt-sensitivity variants. *Am. J. Hum. Genet.* 75(6):1059–1069.

Wagner GP (2001), What is the promise of developmental evolution? Part II: A causal explanation of evolutionary innovations may be impossible. *J. Exp. Zool.* 291(4):305–309.

Wagner GP, Pyle AM (2007), Tinkering with transcription factor proteins: the role of transcription factor adaptation in developmental evolution. *Proc. Novartis Found. Symp.* 284: 116–125.

Whyte LL (1965), *Internal Factors in Evolution*. George Braziller, New York.

Yoffe, Emily (1994), Is Kary Mullis God? Nobel Prize winner's new life. *Esquire* 122(1): 68–75.

6

FROM CELL–CELL COMMUNICATION TO THE EVOLUTION OF INTEGRATED PHYSIOLOGY

Evolutionary Biology, Cell–Cell Communication, and Complex Disease, First Edition.
John S. Torday and Virender K. Rehan.
© 2012 Wiley-Blackwell. Published 2012 by John Wiley & Sons, Inc.

Progress in our knowledge of physiology and medicine is staggering when you consider all of the genomic knowledge and enabling technologies that we now have at our disposal. For example, a recent blue ribbon panel of the American Academy of Arts and Sciences, charged with determining how to ameliorate the crisis in US funding for biomedical research, recommended investing in young scientists and in high-risk, high-reward research (2009). But the problem is far more fundamental than young versus old scientists, or what questions to emphasize. It is due to the lack of an effective and accessible algorithm for readily translating genes into phenotypes—the data operating system.

The malaise in biology as the basic science behind physiology and medicine is even more striking and apparent when compared to the fundamental advances that have been made in physics since Darwin's publication of *The Origin of Species*, starting with Mendeleev's version of the Periodic Table in 1869, which merged chemistry and physics, followed by quantum mechanics and Einstein's formulation of $E = mc^2$. In contrast, Darwin set us off in search of our evolutionary origins at virtually the same time that Mendeleev created the Periodic Table, yet we still don't have a working model for evolutionary biology, or a central theory of biology. That paradox is underscored by publication of the *Human Genome* in 2000, which is lacking an equation for readily converting genes into phenotypes, which would have the impact of $E = mc^2$.

To some extent, the failure to advance medicine is due to the high expectations raised by the Human Genome Project (HGP), and by the promise of systems biology as a readily available and effective means of reconstructing physiology from genes. Like the atom is to physics, the cell is the smallest functional unit of biology. Trying to reassemble gene regulatory networks (GRNs) from acellular extracts of DNA and RNA will result only in a genomic atlas, but not in a mathematical equation or algorithm for interconverting genes and phenotypes. By analogy, grinding up a complex painting like Picasso's *Guernica* in a blender, and then chemically analyzing it would not, for example, convey the tragedy of war and the suffering it inflicts on individuals. Such an analysis requires an integrated, contextual approach.

Indeed, reductionism is a recurrent theme in evolutionary biology, swinging back and forth like a pendulum between genes and phenotypes over its long, stormy, and petulant history. As a result, experimental evidence for the existence of morphogenetic fields like those described by Spemann back in the nineteenth century has been realized only relatively recently. The scientific validity of morphogenetic fields has now been borne out by contemporary molecular embryology, beginning with the breakthrough discovery of homeobox genes, demonstrating the homologies across phyla first proposed by Geoffroy St-Hillaire back in the nineteenth century.

Evolution is emergent and contingent, like the weather or the stock market—neither of which can be predicted on the basis of today's events. Evolutionists have dealt with this complex problem by reducing it to phenotypes and genes, yet, like the representation of the seeming duality of the figure–ground image used in

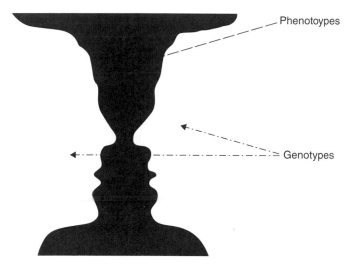

Figure 6.1. *Figure–Ground* is a Gestalt psychology principle first introduced by the Danish phenomenologist Edgar Rubin (1886–1951). Here, we use the duality of the two profiles and the candlestick image to contrast this representation of the apprently dual nature of phenotypes and genotypes in evolutionary biology.

Gestalt psychology (Fig. 6.1), in reality evolution is a process determined by both genes and phenotypes, and natural selection acts on the phenotype to select for the genetic characteristic. The question, then, is: What process integrates the mechanisms for this selection process?

CELL–CELL SIGNALING AND ALVEOLAR DEVELOPMENT: A REDUCTIONIST APPROACH TO THE EVOLUTION OF PHYSIOLOGIC TRAITS

The primordial endodermal and mesodermal germ layers of the developing lung interact over the course of its morphogenesis, ranging in duration from the final 9 days of gestation in a mouse to 222 days in humans, differentiatiating into over 40 different cell types. This process culminates in the production of pulmonary surfactant by the alveoli, which is necessary for preventing the airsacs from collapsing on air breathing. We know a great deal about how growth factor signaling determines these processes, and the downstream signals that alter nuclear expression that ultimately lead to metabolic cooperativity between cells, mediated by ligand–receptor interactions, culminating in alveolar homeostasis.

We have learned much about pulmonary biology since the 1960s by focusing on the production of lung surfactant, the major factor underlying alveolar structure and function. The physiologic significance of lung surfactant was first demonstrated by Avery and Mead (1959), who showed that infants dying of hyaline membrane disease were surfactant-deficient. King and Clements (1972) were the first to chemically define the surfactant as a mixture of phospholipids and proteins. Since then, the basic cellular–molecular biology of the surfactant system has been reduced to a process of communication between the alveolar interstitial lipofibroblast and its neighboring alveolar type II cell (see Chapter 2), which produces both the surfactant phospholipids and proteins. Similarly, Darwin postulated that the prototypic eye consisted of two cells, a photoreceptor and a pigment cell—a process that we now know is controlled by the gene Pax6. The subsequent evolution of the various eye types occurred by adding on to this original genetic program. Gehring has hypothesized a process of intercalary evolution that assumes that the eye morphogenetic pathway is progressively modified by insertion of genes between the master control genes at the top of the hierarchy, and the structural genes such as rhodopsin at the bottom, again raising the question as to how cellular mechanisms caused these evolutionary changes.

Using the simple and practical solution to the problem of the evolution of biosynthetic pathways first suggested by Horowitz (1945), assuming a retrograde mode of biochemical pathway evolution (see Chapter 4, section on reverse engineering, second paragraph), we can trace the evolution of the surfactant all the way back to cholesterol as the key molecular bottleneck, or constraint for the transition from prokaryotes to eukaryotes (see Chapter 1). Konrad Bloch thought that by determining the pathway for cholesterol synthesis, he could trace what he referred to as the molecular fossils that generated cells. He demonstrated that cholesterol evolved on the appearance of oxygen in the atmosphere, and speculated that the biological advantage associated with cholesterol may have been due to the increased fluidity caused by insertion of cholesterol into phospholipid bilayer membranes. Lung surfactant could be regarded as a membrane homolog since it is composed of both neutral lipids such as cholesterol, phospholipids such as lecithin and sphyingomyelin, and apoproteins that facilitate the formation of tubular myelin, the membranous form of the surfactant. The surfactant is functionally homologous to the cell membrane in acting as a physical barrier against exudation of the tissue contents—and, like the effect of cholesterol on the cell membrane enhancing its fluidity, thinning it out for more efficient gas exchange and cytosis, the lung surfactant evolved to facilitate both gas exchange and nutrient metabolism, first through its role in the fish swim bladder for buoyancy (see schematic, in Chapter 7, Fig. 7.5), and subsequently in generating selection pressure for neutral lipid trafficking (see Chapter 2, Fig.2.1).

Even to the naive observer, it is intuitively obvious that there are patterns of size and shape in biology. Darwin was a master at delineating these patterns, and defining a process by which they might have evolved through what he referred to as *descent with modification*, as well as a descriptive mechanism—*natural selection*. Such descriptive metaphors are anachronistic in the age of genomics because they do not provide a way of reducing evolution to its genetic mechanisms, and therefore they do not generate testable, refutable hypotheses at the gene level—and without an understanding of both how, and in adaptation to what, evolution has occurred, we cannot take advantage of evolution's underlying principles, particularly as they apply to human physiology and medicine. This problem arises repeatedly in various ways that are referred to euphemistically as *counterintuitive*, which is an expedient way of dismissing observations that cannot be explained by the prevailing descriptive paradigm. For example, during the course of vertebrate evolution, organ systems coevolved functionally to link lipid metabolism and respiration (alveolar surfactant and gas exchange), photoreception and circadian rhythms (with the pineal as the "third eye"), blood volume control, and erythropoiesis by the kidney, or why the ear ossicles—maleus, incus, stapes—evolved from the jaw bones of fish. These relationships are seemingly counterintuitive in terms of descriptive biology, yet this may be due to the lack of a fundamental mechanistic understanding of the process of evolution. As a corollary, it is estimated that the chemicals composing the human body are worth only a few dollars, yet medical malpractice suits would indicate quite the opposite, averaging hundreds of thousands, to millions of dollars in settlement charges. Or the evolutionist's frequent conclusion that a certain evolved trait appears to be too costly, until you consider that the alternative to the adaptation in question would have been extinction! These counterintuitive observations are all the result of reasoning after the fact.

Alternatively, to meet the great challenge of implementing genomics to understand biology, we have reconsidered the process of evolution from a cellular–molecular signaling perspective, since that is the mechanistic basis for the formation of complex structures and functions. It seems intuitively obvious that there must be fundamental commonalities between ontogeny and phylogeny, given that both originate from single-celled organisms, forming progressively more complex structures through cell–cell interactions mediated by growth factors and their receptors. By systematically focusing on such cellular–molecular developmental mechanisms as serial events across vertebrate classes, we can test the inferred direction and magnitude of evolutionary change such as those depicted in cladograms using falsifiable hypotheses based on descent with modification.

AN INTEGRATED, EMPIRIC, MIDDLE–OUT APPROACH TO PHYSIOLOGY

The greatest challenge in the postgenomic era is to effectively integrate functionally relevant genomic data in order to derive physiologic first principles, and determine how to use them to decode complex biologic traits. Currently, this problem is being addressed statistically by analyzing large datasets to identify genes that are *associated* with structural and functional phenotypes—whether they are in fact causal seems to be inconsequential. This approach is merely an extrapolation from biological classifications, beginning with Linnaeus' binomial

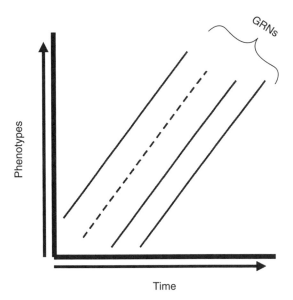

Figure 6.2. Solving for evolutionary principles independently of chronologic time. By regressing gene regulatory networks (GRNs) mediating cell–cell interactions relevant to structure and function across ontogeny and phylogeny against chronologic time and phenotype, we can generate a family of parallel lines, or simultaneous equations. Using this approach, we can solve for the underlying evolutionary principles involved, independently of chronologic time, making the biologic processes self-referential.

nomenclature. The reductionist genetic approach cannot simply be computed to generate phenotypes—evolution is not a result of chance; it is an *emergent and contingent* process. In both the current and future research environments, we must expand our computational models to encompass the broader evolutionary approach.

Earlier in this book, we formally proposed using a comparative, functional genomic, middle–out approach to solve for the evolution of physiologic traits (Chapter 2). This approach engenders development, homeostasis, and regeneration, represented by a family of parallel lines that can be mathematically expressed as a set of simultaneous equations, schematized in Figure 6.2. This perspective provides a *testable and refutable* way of systematically integrating such information in its most robust form to retrace its evolutionary origins. For example, as depicted in Figure 6.2, a GRN common to the phenotypes for development, homeostasis, repair, and aging of a given structure/function (lung, kidney, liver, brain, etc.) can be depicted as changing over chronologic time (x axis). Such a set of simultaneous equations can then be solved for these GRN–phenotype interrelationships in biologic spacetime, or evolution, independent of chronologic time; in other words, all biologic processes are now relative to one another, *independent of chronologic time*. Such a self-referential property of evolved structure and function reflects the modular nature of the cell–cell interaction principle. Such a result would be expected of a novel functional reduction of evolutionary biology. But this primary mechanism of evolution is complicated by the fact that the selection is for genes in specific cell populations as they relate to specific physiologic functions, such as breathing, locomotion, digestion, fluid and electrolyte balance, and cognition, to name only a few primary examples. But those same genes are expressed in all of the cells in that cell lineage, both with regard to the primary structure and function being selected for, and for all of the other tissues and organs where that cell type is present. The descriptive term for this phenomenon is *exaptation* (preadaptation), a term coined by Gould and Vrba (1982). An exaptation is defined as either a character previously shaped by natural selection for a particular function (an adaptation), which is then coopted for a new use, or as a character whose origin cannot be ascribed to the direct action of natural selection, which is coopted for a current use. What Gould and Vrba had not considered were the developmental consequences of such a process. It would create scenarios in which cells of differing lineages would be forced into spatiotemporal juxtapositions based on developmental principles, whereas the formation of novel gene regulatory networks would either create novel structures and/or functions, or not, depending on whether they were compatible with viability, or were ultimately constrained by the reproductive process.

Among mammals, embryonic lung development is subdivided into two major phases: branching of the airways and the formation of alveoli. Fortuitously, we have observed that deleting the parathyroid hormone–related protein (PTHrP) gene specifically results in failed alveolization (Rubin et al. 1994)—the generation of progressively smaller, more numerous alveoli with thinner walls for efficient gas exchange was necessary for the transition from water to air, and for the further evolution of land vertebrates. This, and the fact that PTHrP and its receptor are highly conserved (the PTHrP ortholog tuberoinfundibular protein (TIP39) is expressed as far back in phylogeny as yeast), is stretch-regulated, and forms a paracrine signaling pathway that mechanistically links the endodermal and mesodermal germ layers of the embryo to the blood vessels has compelled us to exploit this key transitional GRN to gain insight to the first principles of respiratory physiology. This model can transcend pulmonary physiology by extrapolating the adaptive cell–cell signaling mechanisms of the lung using ontogenetic and phylogenetic principles, as follows. PTHrP produced by the lung epithelium regulates mesodermal leptin production through a receptor-mediated mechanism. We have implicated leptin in the normal paracrine development of the lung, demonstrating its effect on lung development in the *Xenopus laevis* tadpole (Torday et al. 2009), for the first time providing a functional, cellular–molecular mechanism for the frequently described coevolution of metabolism, locomotion, and respiration as the basis for vertebrate evolution. Leptin acts as an organizing principle for PTHrP/PTHrP receptor-mediated alveolar homeostasis (please refer to Chapter 2)—leptin is a ubiquitous product of fat cells, which binds to its receptor on the alveolar epithelium of the lung, stimulating surfactant synthesis, reducing surface tension, and generating an increasingly compliant structure on which natural selection could ultimately select for the stretch-regulated PTHrP coregulation of surfactant and microvascular perfusion. This mechanism may ultimately have given rise to the contemporary mammalian lung, with its maximal surface area for gas exchange functionally coupled with both stretch-regulated surfactant production and alveolar capillary perfusion. These adaptations were the direct result of the thinner alveolar walls due to PTHrP's apoptotic, or *programmed cell death*, effect on fibroblasts, and a structurally reinforced blood–gas barrier due to the evolution of type IV collagen (West and Mathieu-Costello 1999). This last feature may have contributed more generally to the molecular body plan based on the coevolution of type IV collagen in the lung alveoli and kidney glomeruli, as follows.

A MOLECULAR EVOLUTIONARY LINK BETWEEN THE LUNG AND THE KIDNEY?

In Chapter 5, we discussed the role of type IV collagen in the evolution of the lung, based on its high tensile strength compared to other collagen isotypes. The significance of type IV collagen in our understanding of the evolution of integrated physiology transcends its physical strength, since its physicochemical composition appears to have facilitated the evolution of both the lung alveolus and the kidney glomerulus. This evolutionary perspective on type IV collagen in the alveolus and glomerulus is underpinned by other physiologic commonalities between these structures and their functions. In that context, it is noteworthy that both the lung and the kidney produce fluid in the womb in mammals. Furthermore, both alveoli and glomeruli are *stretch sensors* (see Fig. 6.3). The epithelial type II cells sense the stretching of the alveolar wall by producing PTHrP, which stimulates the neighboring fibroblasts to produce leptin. Leptin then stimulates surfactant production by the type II cells. Similarly, the glomerular mesangium, a modified fibroblast that senses and monitors the pressure created by the flow of blood through the glomerulus, regulates the urinary output of the kidney. PTHrP produced by podocytes within the glomerular wall maintains the physiologic stretch-sensing function of the mesangium. In both the alveolus and glomerulus, failure of PTHrP signaling causes transdifferentiation of resident fibroblasts to myofibroblasts, causing fibrosis, leading to failed physiologic responses to stretching in both cases. Therefore, from this cellular–molecular evolutionary perspective, both alveolar and glomerular physiology are homologous with respect to both their functional adaptation to stretch and their seemingly maladaptive fibrotic responses to disease. Viewed from an evolutionary perspective, the ability to function until the organism is able to reproduce would be adaptive. Such a molecular integration based on fundamental physiologic principles provides

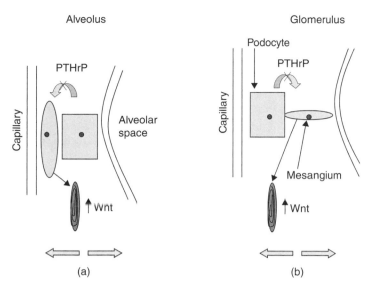

Figure 6.3. The alveolus and glomerulus are stretch sensors. In the lung (a), the alveolar epithelium (square) and fibroblast (oval) respond to the stretching of the alveolar wall by increasing surfactant production. In the kidney (b), the mesangium (oval) senses fluid pressure and regulates bloodflow in the glomeruli. In both cases, breakdown in cell–cell interactions causes these cells to become fibrotic (brown cell) due to upregulation of Wnt. (See insert for color representation.)

important molecular homeostatic insights to the physiology and pathophysiology of the lung and the kidney.

THE BERNER HYPOTHESIS AND EMERGENCE OF THE ADIPOCYTE: THE EVOLUTIONARY ORIGINS OF THE LIPOFIBROBLAST

Given the central role of the lipofibroblast in vertebrate lung evolution, both as a cytoprotective mechanism against oxygen-free radicals and as an integrator of stretch-regulated surfactant production, one might ask how and why it might have evolved. Csete et al. (2001) observed that cultured muscle stem cells spontaneously differentiate into fat cells in 21% oxygen (room air), but not in 6% oxygen. This experimental observation is provocative in light of the periodic rises and falls in atmospheric oxygen over the past 500 million years, as shown by Berner et al., (2007) suggesting that the increases in atmospheric oxygen may have induced the formation of adipocytes over the course of vertebrate evolution. Our laboratory has shown that the muscle-derived adipocyte-like lipofibfroblasts in the lung protect the alveoli against oxidant injury, and promote surfactant synthesis by producing leptin. The coevolution of these features would have facilitated the vertebrate transition from water to land, and may have fueled the progressive increase in oxygenation from amphibians to reptiles and mammals by increasing alveolar distensibility. Also, the oxygen stimulus for lung evolution may have been complemented by the accompanying decreases in atmospheric oxygen, since hypoxia is the most potent physiologic stimulus for adrenaline production. Thus, the episodic fluctuations in atmospheric oxygen would have reinforced the selection pressure for the β-adrenergic receptor mechanism for alveolar homeostasis (see Chapter 5, Fig. 5.1), including surfactant production, fluid balance, and microcirculatory blood pressure regulation.

A natural consequence of the increase in tissue oxygenation resulting from the evolution of the lung is the differentiation of peripheral muscle stem cells into fat cells (Csete et al. 2001). This may also have led to the evolution of endothermy, or warm-bloodedness, which requires increased metabolic activity, putting further selection pressure on the cardiopulmonary system. The increase in body temperature from cold- (25°C) to warm-blooded organisms (37°C) in and of itself would have increased surfactant phospholipid activity by 300%. For example, map turtles (Graptemys geographica) show different surfactant compositions depending on the ambient temperature. Therefore, the advent of thermogenesis would have been facilitated by the physical increase in lung surfactant surface tension-lowering activity. These synergistic selection pressures would have been further enhanced directly by the coordinate physiologic effects of epinephrine on the heart, lung, and fat depots, and indirectly by the increased production of leptin by fat cells, which is known to promote the formation of blood vessels and bone, accommodating the infrastructural changes necessitated by the evolution of complex physiologic traits. Crespi and Denver (2006) have shown that leptin stimulates both food intake and limb development in *Xenopus laevis* tadpoles, functionally linking metabolism and locomotion through this pleiotropic mechanism. Importantly, we subsequently discovered that leptin also stimulates lung development in these tadpoles (Torday et al. 2009), now functionally linking metabolism, locomotion, and respiration—the three major functional characteristics of vertebrate evolution—together mechanistically for the first time through these pleiotropic effects of leptin.

This cellular–molecular functional genomic connection between molecular oxygen, adipogenesis, leptin, and the evolution of land vertebrates is reminiscent of Theseus following the string to escape from the labyrinth! Evolutionary biology needs more such experimental models based on refutable hypotheses to solve the evolutionary puzzle–namely, the systematic identification of deep homologic physiologic traits, which, when reduced to those cellular–molecular features that have been modified by descent, can be experimentally tested. For example, we are expanding our experiments on the effects of leptin on lung development in *Xenopus* to determine how it affects the blood–gas barrier basement membrane and vasculature. In the future, we will determine whether the same or similar cellular–molecular mechanisms apply to fish, reptiles, birds, and mammals. The take-home message is that the elucidation of the mechanistic origins of molecular homologies is an experimentally soluble problem.

LUNG BIOLOGY AS A CIPHER FOR EVOLUTION

The lung is an on-demand system, in which the production of surfactant, alveolar capillary bloodflow and type IV collagen synthesis are all under the control of the PTHrP-leptin stretch mechanism. Why? In theory, these processes could have functioned at full capacity due to neutral theory or natural selection, as suggested by the symmorphosis hypothesis (Weibel et al, 1991). Instead, the mechanisms are now, and were then, functionally adapted to the prevailing environmental conditions, which were revealed to be perhaps essentially exorbitant on conventional *ex post facto* physiologic analyses. Yet, when seen prospectively from a cell evolutionary perspective, as an epistatic mechanism, such critically important adaptations, in the aggregate, ultimately represent an all-or-nothing, life-or-death survival-of-the-fittest event. When viewed from that vantage point, the evolution of the lung from the swim bladder now provides a rational explanation for the on-demand mechanisms of lung physiology because the expansion and contraction of the swim bladder with air as an adaptation to buoyancy (i.e., gravity) was the basic bauplan. It provided the physiologic basis for the emergence of the alveolus as the organ of gas exchange. As a note in support of this concept, when Weibel and his colleagues tested the symmorphosis principle in the lung, they came to the conclusion that it was "overengineered," that is, that it had far more capacity than was required by physiologic criteria. That may appear to be the case only according to the assumption that the lung evolved under selection pressure for gas exchange. Alternatively, if the lung initially evolved as an outpouching of the gut in adaptation to buoyancy and efficient feeding, the primary selection pressure may not have been for oxygenation, but for traits relevant to gut barrier function. Given that surfactant proteins A and D are innate host defense mechanisms, it is feasible that the primordial selection pressure for the evolving lung was for host defense, followed by selection pressure for oxygenation through mechanisms of reduced surface tension. Also, the algae that fish feed on at the surface are 90% lipid, providing a common selection pressure for the swim bladder as the prototype for lung evolution and for lipid metabolism. Therefore, by considering the physiologic origins of the lung on the basis of deep homologies rather than on superficial, contemporary functional principles, the underlying evolved principles of lung physiology can be accessed. Ultimately, such analyses will generate an evolution-based logic for physiology that will be become intuitive and predictive, rather than counterintuitive and postdictive.

DO STRETCH EFFECTS ON PTHrP EXPRESSION REFLECT ITS ROLE IN ADAPTING TO GRAVITY?

Both lung and bone share a common requirement for tension to maintain their functionality. Since PTHrP is an interactive paracrine factor that is necessary for

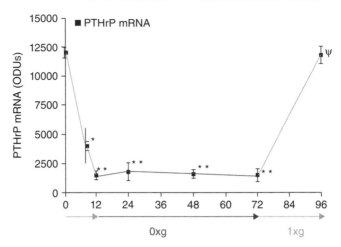

Figure 6.4. Human UMR 106 bone cells were maintained in a rotating wall vessel for ≤72 h. At the end of that time, the cells were returned to unit gravity (1 × g) for 24 h. Cells were analyzed for PTHrP mRNA expression using reverse transcriptase polymerase chain reaction (RT-PCR). [*Notation:* $N = 6$; *, $p < 0.00001$; **, $p < 0.000001$; ψ, $p > 0.05$ vs. time = 0 by analysis of variance. The units on the y axis (ordinate) labeled ODU stand for optical density units.] (See insert for color representation.)

the homeostatic control of both bone and lung, and stretch affects PTHrP expression in both, it was hypothesized that loss of tension on bone or lung cells, both of which express PTHrP, would cause a decrease in PTHrP expression (previously alluded to in Chapter 4, beginning of section on the cell–cell communication model of lung evolution). To test that hypothesis, fetal rat lung alveolar type II cells were attached to floating beads and suspended in freefall using a rotating wall vessel to mimick zero gravity (0 × g) (see Fig. 6.4). Cells were harvested over time, and analyzed for PTHrP mRNA expression. PTHrP messenger RNA (mRNA) expression decreased significantly over an 8–12-h period, reaching a stable nadir for 12–72 h. When the cells were returned to unit gravity, PTHrP mRNA expression was fully recovered within 24 h, suggesting that the inhibitory effect of 0 g is specifically due to the effect of gravity.

This effect of gravity on PTHrP expression was subsequently evaluated *in vivo* (Torday 2003). Bones from rats flown in 0 g were obtained from the NASA Spacelab Life Sciences 2/STS-58 mission, which was launched on the space shuttle *Columbia* on October 18, 1993. After 14 days in orbit, one of the longest missions in US manned space history to date ended on November 1, 1993, when *Columbia* landed at Edwards Air Force Base, California.

The primary mission objective was to conduct experiments within the Spacelab Life Sciences 2 (SLS-2) payload, an array of life science investigations using the laboratory facilities housed in the Spacelab module. The six experiments spon-sored by the Ames Research Center used rats as research subjects. Experience gained from the SLS-1 mission aided in the operational aspects of SLS-2. On this mission, for the first time in the history of US spaceflight, the crew conducted blood draws and tissue dissections in flight. Conducting such procedures in space enabled scientists to clearly distinguish between the effects of microgravity and the effects of landing and readaptation to Earth gravity. Rat tissues collected in flight were preserved and distributed to scientists from the United States, Russia, France, and Japan through an extensive biospecimen sharing program.

The goal of the SLS-2 mission was to study the structural and functional changes occurring in the bone, muscle, blood, and balance systems of rats and humans in response to spaceflight. The SLS-2 experiments were intended to supplement data gathered on these bodily changes during previous US payloads and Soviet/Russian missions, including SL-3, Cosmos 1667, 1887, and 2044, prior to 1991, and SLS-1 in 1991. Bone experiments were designed to study the changes in calcium metabolism, bone formation, and mineralization that occur in micro-gravity. Muscle studies focused on microgravity-induced atrophy. Hematology experiments examined red blood cell shape and levels, blood cell mass, plasma volume, and blood cell formation. Neurophysiology studies examined the struc-ture of gravity receptors and the physiologic changes that may be involved in the space adaptation syndrome. In previous experiments, scientists assessed spaceflight-induced physiologic changes by dissecting and analyzing animal tissues several hours after the subjects had returned to Earth. Postflight dissection did not permit researchers to clearly differentiate between the effects of spaceflight itself on the organisms and the effects of readaptation to Earth gravity, since readaptation to Earth gravity occurs rapidly in some systems and tissues. The SLS-2 experiments were the first spaceflight experiments to assess changes occurring in tissues while organisms were still in space.

The experiments used male, pathogen-free white albino rats (*Rattus norvegicus*) of the Sprague–Dawley strain. Flying a large contingent of rats allowed investiga-tors to gather statistically significant data for a number of parameters. Ground control groups used 106 rats. Each of the flight rats weighed approximately 300 g at launch.

The space- and ground-based rat bones were analyzed for their PTHrP mRNA content. There were significant decreases in PTHrP mRNA expression in the tibias and femurs of space-based compared to ground-based rats (see Fig. 6.5). However, in contrast to this observation, there were no differences in PTHrP mRNA expres-sion by the skull bones. These findings are consistent with the hypothesized effect of gravity on PTHrP mRNA expression, since it is generally thought that such unweighting effects are due to tension transmitted to weight-bearing bones by the muscle, ligaments, and tendons—which would not have affected the skull bones.

Knowledge of the effects of gravity on humans derives from the physiologic study of astronauts. There is a well-documented linear loss of bone among

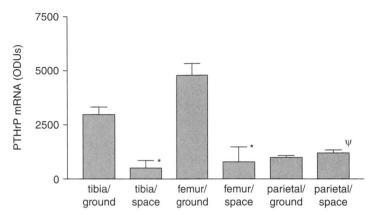

Figure 6.5. PTHrP mRNA expression by bones of rats flown in space for 2 weeks were compared to ground-based littermate rat bones. (*Notation:* $N = 5$; *, $p < 0.05$; ψ, $p > 0.05$ vs. ground control, by Student's t test; ODU = optical density unit.)

astronauts as a function of the length of time they spend in space, suggesting that microgravity somehow causes the loss of calcium from bone. That process is consistent with our experimental evidence for a microgravitational effect on PTHrP expression by osteoblasts, since PTHrP controls bone calcification.

West et al. (1997) have also shown that microgravity affects lung function by disrupting the normal regional distribution of alveolar gas exchange from the apex to the base of the lung. We speculate that this, too, may have resulted from altered PTHrP expression, as a result of the loss of PTHrP regulation of ventilation perfusion (V/Q) matching, since PTHrP integrates alveolar distension, surfactant production, and alveolar capillary perfusion, or (V/Q) matching, which would be predicted to affect regional distribution of oxygen in the lung.

From an evolutionary perspective, these observations are consistent with PTHrP mediating the adaptational response to gravity; the lung is structurally derived from the lower pharynx, and functions to adapt to gravity by moving gases (O_2, out of the blood into the swim bladder to correct for buoyancy—i.e., it is an *anti-gravity* device).

WOLFF'S LAW WORKS FOR BOTH BONE AND LUNG

Bone tissue can adapt to its functional environment to optimize its mechanical demand. Wolff's law, proposed in 1892, was the first to suggest that the skeleton could respond to such functional demands. It is the cells of the skeleton—the osteoblasts, osteocytes, and osteoclasts—that mediate the remodeling process. Investigators have focused on how strain affects signaling to regulate bone cells. Cellular regulation of bone remodeling involves both the sensing and response to the mechanical environment. Exercise studies reveal that certain degrees of loading must be exceeded to see changes, and that the degree of the response is determined not only by the magnitude of the load but also by the rate, cycle number, and even the frequency of the load. Bone loss during flight in deep space causes a dramatic loss in bone tissue, approaching 2% bone density loss per month. Disuse can also cause a rapid loss of bone similar to that seen during spaceflight. These observations of gain and loss of bone under varying mechanical loads indicates that mechanosensitivity plays a role in regulating bone density.

FUNCTIONAL RELATIONSHIP BETWEEN THE EXTERNAL AND INTERNAL ENVIRONMENTS

The foregut is a plastic structure from which the thyroid, lung, and pituitary arise through Nkx2.1/TTF-1 expression. Evolutionarily, this is consistent with the concept of *terminal addition*, since the deuterostome gut is formed from the anus to the mouth. Moreover, when Nkx2.1/TTF-1 is deleted in embryonic mice, the thyroid, lung, and pituitary do not form during embryogenesis, providing experimental evidence for the genetic commonality of all three organs. Their phylogenetic relationship has been traced back to amphioxus, and to cyclostomes, since the larval endostyle, the structural homolog of the thyroid gland, expresses Nkx2.1/TTF-1.

The Phylogeny of the Thyroid

The endostyle is retained in postmetamorphic urochordates, and in adult amphioxus, but the postmetamorphic lamprey has a follicular thyroid gland, which is a transformed endostyle. The presence of an endostyle in larval lampreys does not suggest direct descent of lampreys from protochordates, but instead that the evolutionary history of lampreys is long and of ancient origin, and that they share the common feature of having filter-feeding mechanisms in their larval stage of development. However, it is noteworthy that the other extant agnathan, the hagfish, possesses thyroid follicles before hatching. Since hagfish evolution is considered to be conservative, and hagfish history can be traced back ~550 million years, this suggests that thyroid follicles could likewise be considered to have an ancient history.

Differences in the ontogeny of the thyroid gland is another example of early divergence of lampreys and hagfish during their evolutionary history. On the other hand, it is of interest that the method of development of thyroid follicles from a broad pharyngeal epithelium in hagfish embryos is similar to that seen during lamprey metamorphosis, when follicles arise from clumps of cells emanating from the transforming endostyle epithelium. Perhaps hagfish embryology reflects a step in the development of agnathan thyroid follicles that occurred later in lampreys, when metamorphosis appeared in their ontogeny.

Eales (1997) has reviewed the phylogenetic history of the endostyle and nonfollicular thyroid tissues of vertebrates and invertebrates. He emphasized the importance of exogenous compounds and peripheral mechanisms in regulating thyroidal status, and in influencing the course of the phylogenetic development of the vertebrate thyroid gland. A key to thyroidlike function in a tissue is the ability to salvage iodine, and there are many nonchordate invertebrates that have iodine-binding ability. Ingested and or absorbed free iodocompounds, perhaps even iodothyronines from plants and microorganisms, can be metabolized. Metabolism in the gut may take place with the assistance of resident bacteria. Absorption of iodocompounds has not been shown in the larval lamprey intestine, but this organ is the primary site for monodeiodination from T4 to T3, and this also seems to be the case for ascidians. The endostyle is found only in marine invertebrates with a notochord, and in the freshwater-dwelling larval lampreys. Although this specialized region of the alimentary tract may have evolved following selection pressure favoring filter feeding, like many tissues of invertebrates it had iodine-binding capacity, a function that developed secondarily. The iodine-binding capacity of endostyle in extant ascidians and larval lampreys is well documented. The retention of the endostyle throughout the evolutionary history of lampreys is a reflection of its importance in the filter-feeding apparatus, and also suggests that it may have appeared when lampreys were in a pelagic marine phase.

The vertebrate follicular thyroid gland evolved following strong selection pressure for a gland that favored thyroid hormone and iodine storage during the period when ancient chordates moved from the iodine-rich marine environment to the iodine-poor freshwater habitat. In this interpretation, the presence of an endostyle in extant larval lampreys is a reflection of their ancient marine origins, and the adult follicular thyroid gland came after the animal moved toward freshwater. Since the follicular thyroid gland of adult lampreys appears only following transformation of the endostyle during metamorphosis, there was selection pressure created by the freshwater environment for metamorphosis in the ontogeny of lampreys. A consequence of this metamorphosis was transformation of the endostyle into a follicular thyroid gland.

This view of the ontogeny of the lamprey thyroid gland, in conjunction with the belief that the freshwater habitat of lampreys is a secondarily acquired niche, implies that metamorphosis might not have originated as a developmental strategy, but occurred when lampreys moved into freshwater during their evolutionary history. An alternate view would be that the presence of metamorphosis, and ultimately the follicular thyroid gland, in the marine environment was the primary reason why lampreys could inhabit freshwater. The presence of an endostyle in early lampreys was not a requisite for living in a marine (brackish) environment, since present-day larvae with endostyles cannot tolerate even dilute seawater. Premetamorphic lampreys have lost the ability of the presumed (hypothetical) ancestral larval form to inhabit a marine environment of any type. It is only after metamorphosis that juveniles of some species can tolerate full-strength seawater.

Larval and reproductive intervals of the lamprey lifecycle are confined to freshwater. The ability of juveniles of some species to osmoregulate in seawater could also be a character that was secondarily derived following the appearance of metamorphosis in the ontogeny of lampreys. Extant parasitic lampreys of the most ancient lineage (e.g., *Ichthyomyzon unicupis*) are confined to freshwater. An explanation of how marine osmoregulation may have been secondarily derived following the primary derivation of metamorphosis may be found in modern views of the phenomenon called *developmental integration.*

AN EVOLUTIONARY VERTICAL INTEGRATION OF THE PHYLOGENY AND ONTOGENY OF THE THYROID

Mechanistically, the increased bacterial load due to the facilitation of feeding by the endostyle may have stimulated the cyclic AMP-dependent protein kinase A, or PKA, pathway, since bacteria produce endotoxin, a potent PKA agonist. This cascade may have evolved into regulation of the thyroid by thyroid stimulating hormone (TSH), since TSH acts on the thyroid via the cAMP-dependent PKA signaling pathway. This mechanism potentially generated novel structures such as the thyroid, lung, and pituitary, all of which are induced by the PKA-sensitive Nkx2.1/TTF-1 pathway. The brain–lung–thyroid syndrome (Carré et al. 2009), in which infants with Nkx2.1/TTF-1 mutations develop hypotonia, hypothyroidism, and respiratory distress syndrome, or surfactant deficiency disease, is further evidence for the coevolution of the lung, thyroid, and pituitary.

Developmentally, the thyroid evaginates from the foregut in the mouse beginning on day 8.5, one day before the lung and pituitary emerge, suggesting that the thyroid may have been a molecular prototype of the lung during evolution, providing a testable and refutable hypothesis. The thyroid rendered molecular iodine in the environment bioavailable by binding it to threonine to generate thyroid hormone, whereas the lung made molecular oxygen bioavailable, first by inducing fat cell–like lipofibroblasts as cytoprotectants, which then stimulated surfactant production by producing leptin, placing increased selection pressure on the blood–gas barrier by making the alveoli more compliant. This, in turn, may have created further selection pressure for the metabolic system to utilize the rising oxygen level in the environment, placing further selection pressure on the alveoli, giving rise to the stretch-regulated surfactant system mediated by PTHrP and leptin. Subsequent selection pressure on the cardiopulmonary system may have facilitated liver evolution, since the progressively increasing size of the heart may have induced precocious liver development, fostering increased glucose regulation. The brain serves as a glucose sink, and there is experimental evidence that increasing glucose during pregnancy increases the size of the developing brain (Saintonge and Coté 1984). The further evolution of the brain, specifically the pituitary, would have served to further the evolution of complex physiologic systems.

Both the thyroid and the lung have played similar adaptive roles during vertebrate evolution. The thyroid has facilitated the utility of iodine ingested from the environment, whereas the lung has accommodated the rising oxygen levels during the Phanerozoic era. In both cases, these structures have accommodated otherwise toxic substances for biologic purposes that have allowed vertebrates to adapt to their environment. Importantly, the thyroid and the lung may have interacted cooperatively in facilitating vertebrate evolution. Thyroid hormone stimulates embryonic lung morphogenesis during development, while also accommodating the increased lipid metabolism needed for surfactant production by driving fatty acids into muscle to increase motility, as opposed to oxidization of circulating lipids to toxic lipoperoxides. The selection pressure for metabolism was clearly facilitated by the synergy between these foregut derivatives.

The lingering question is how the lung could have formed the mechanistic basis for the evolution of other organs and physiologic systems. One potential mechanism is that the selection pressure for the cumulative lung *cis* regulatory adaptation mechanisms favored those individuals with those particular molecular traits, leading to other cis regulatory adaptations—what some are calling *evolvability*. In Chapter 7 we will discuss how the cell–cell signaling model of physiologic evolution may have evolved through selection pressure, and compare this mechanistic, predictive approach to the conventional descriptive approach to evolution.

REFERENCES

Avery ME, Mead J (1959), Surface properties in relation to atelectasis and hyaline membrane disease. *AMA J. Dis. Child.* 97(5, Pt.):517–523.

Berner RA, Vandenbrooks JM, Ward PD (2007), Evolution—oxygen and evolution. *Science* 316:557–558.

Carré A, Szinnai G, Castanet M, Sura-Trueba S, Tron E, Broutin-L'Hermite I, Barat P, Goizet C, Lacombe D, Moutard ML, Raybaud C, Raynaud-Ravni C, Romana S, Ythier H, Léger J, Polak M (2009), Five new TTF1/NKX2.1 mutations in brain-lung-thyroid syndrome: Rescue by PAX8 synergism in one case. *Hum. Mol. Genet.* 18(12):2266–2276.

Crespi EJ, Denver RJ (2006), Leptin (ob gene) of the South African clawed frog Xenopus laevis. *Proc. Natl. Acad. Sci. USA* 103(26):10092–10097.

Csete M, Walikonis J, Slawny N, Wei Y, Korsnes S, Doyle JC, Wold B (2001), Oxygen-mediated regulation of skeletal muscle satellite cell proliferation and adipogenesis in culture. *J. Cell Physiol.* 189(2):189–196.

Eales JG (1997), Iodine metabolism and thyroid-related functions in organisms lacking thyroid follicles: are thyroid hormones also vitamins? *Proc. Soc. Exp. Biol. Med.* 214(4):302–317.

Gould SJ, Vrba ES (1982), Exaptation—a missing term in the science of form. *Paleobiology* 8(1):4–15.

Horowitz NH (1945), On the evolution of biochemical syntheses. *Proc. Natl. Acad. Sci. USA* 31(6):153–157.

King RJ, Clements JA (1972), Surface active materials from dog lung. II. Composition and physiological correlations. *Am. J. Physiol.* 223(3):715–726.

Rubin LP, Kifor O, Hua J, Brown EM, Torday JS (1994), Parathyroid hormone (PTH) and PTH-related protein stimulate surfactant phospholipid synthesis in rat fetal lung, apparently by a mesenchymal-epithelial mechanism. *Biochim. Biophys. Acta.* 1223(1):91–100.

Saintonge J, Côté R (1984), Fetal brain development in diabetic guinea pigs. *Pediatr. Res.* 18(7):650–653.

Torday JS (2003), Parathyroid hormone-related protein is a gravisensor in lung and bone cell biology. *Adv. Space Res.* 32(8):1569–1576.

Torday JS, Ihida-Stansbury K, Rehan VK (2009), Leptin stimulates Xenopus lung development: Evolution in a dish. *Evol. Devel.* 11(2):219–224.

Weibel ER, Taylor CR, Hoppeler H (1991), The concept of symmorphosis: A testable hypothesis of structure-function relationship. *Proc. Natl. Acad. Sci. USA* 88(22):10357–10361.

West JB, Mathieu-Costello O (1999), Structure, strength, failure, and remodeling of the pulmonary blood-gas barrier. *Annu. Rev. Physiol.* 61:543–572.

West JB, Elliott AR, Guy HJ, Prisk GK (1997), Pulmonary function in space. *JAMA* 277(24):1957–1961.

7

EXPLOITING CELL–CELL COMMUNICATION ACROSS SPACETIME TO DECONSTRUCT EVOLUTION

We shall not cease from exploration
And the end of all our exploring
Will be to arrive at what we started
And know the place for the first time
—T. S. Eliot, Four Quartets

Darwin has been called the Newton of biology, but it will be time enough to talk about the Newton of biology after our science has found Galileo.
—J. H. Woodger, 1929

Evolutionary Biology, Cell–Cell Communication, and Complex Disease, First Edition.
John S. Torday and Virender K. Rehan.
© 2012 Wiley-Blackwell. Published 2012 by John Wiley & Sons, Inc.

Evolution is a process, so why do we tend to dwell on phenotypes and genes instead? The problem with selecting a process is, which one? By looking at the various processes of biology—development, physiology, regeneration, and aging—individually, we only see glimpses of evolution, but not the entire process as an integrated whole, like the tale of the elephant and the six blind men, each with his own interpretation. Yet the process of evolution must lie somewhere between the gene and the phenotype. During development, that space is occupied by cell–cell signaling mechanisms mediated by soluble growth factors. It is that process that we will focus on in this chapter. Is cell–cell signaling the basis for evolution? That is the premise of this chapter.

Biologists have been puzzling over the evolutionary origins of life for the past 150 years, ever since the challenge created by the publication of *The Origin of Species*. Darwin delineated the problem and provided a metaphoric mechanism—natural selection—and we have been attempting to understand how descent occurs with modification to generate "forms most beautiful" ever since. Initially, the embryologists took the lead in the nineteenth century, trying to reconcile development with phylogeny. Starting at the turn of the twentieth century, though, that effort was usurped by the geneticists, for lack of empiric evidence from the embryologists. For example, the developmental organizing principle described by Spemann could not be isolated. The genetic approach was then merged with Darwinian evolution in the modern synthesis in the first third of the twentieth century in a concerted effort to integrate these concepts. More recently, the role of development in evolution has reemerged as evolutionary–developmental biology, or evo-devo. As a result of these varied approaches, evolutionary biology has had a long and embattled history that tends to overshadow the scientific effort to understand the evolutionary origins and first principles of biology. Nowhere is this more evident than in the writings of Conrad Waddington, who in his book *The Evolution of an Evolutionist*, freely admits that evolution is seen in the context of a "uniform environment" as "natural selection exerting itself to specify one single coefficient for each biologic entity"—and yet he acknowledges that phenocopies, or variations in phenotypes, exist as a result of the mechanisms of evolution. Elsewhere, he alludes to the reduction of phenotypes to DNA; this is misleading because we know that genetic traits are determined by genes acting through cell types. In his book *The Strategy of the Genes*, Waddington states that biology must be seen as three types of temporal change: large-scale evolution, medium-scale life history, and short-scale physiology. Waddington also states that

> The science that centres on the short time-scale is physiology. But we have to remember that the processes of day-to-day life do not just keep the animal alive in a constant condition. Repeating day after day, they alter slowly as time passes, and produce the medium scale changes of development. Somewhere hidden among the deepest secrets of the physiology of the cell, there must be the process by which the hereditary factors undergo those sudden mutations which are the basis of the long time-scale evolution.

In this chapter, we will attempt to show that chronologic time is not only unnecessary for understanding the mechanisms of evolution but also actually obscures our vision of the evolutionary process. Evolutionary biology is self-referential, and therefore time is superfluous and misleading once you understand the principles involved.

Because of the historic emphasis on scientific reductionism, the process of evolution and its mechanistic basis have been distilled down to genes and phenotypes, leaving us with associations between the former and the latter. What we need is a way of interrelating genes and phenotypes functionally in their historic contexts in order to deconvolute the evolutionary strategy. Molecular cell embryology achieves that goal by tracing development from the fertilized egg to the multicellular organism, and back again to the unicellular zygote, in reproduction, iteratively, from one lifecycle to the next. This body of information has given us insight into how the embryo is formed, grows, differentiates, and adapts to its environment. We now know that the communication between cells in the developing conceptus is mediated by soluble growth factors, such as the Spemann organizer, that bind to their receptors on neighboring cell types to signal their presence and level of differentiation. The target cell binds growth factors from its cell neighbors, triggering its developmental and resultant homeostatic programs. These growth factors are highly evolutionarily conserved, and are expressed all the way back to the unicellular origins of multicellular organisms, giving us the opportunity to trace the molecular origins of vertebrates. This approach also leads to an understanding of how these cell–cell communications determine physiology, and why the breakdown in cell–cell communication leads to adaptive or maladaptive strategies. This cell's eye view of evolution predicts the aging process, based on the concept that aging is literally development in reverse, since development forms cellular communications, whereas aging is the breakdown in cellular communications (see Chapter 1). Moreover, if the function of the evolutionary process is to mediate the adaptation of species over generations to an ever-changing environment, then reproduction may be seen as the intergenerational optimization of the communication of such knowledge. That precept is important because it provides a rationale for understanding the aging process—if biologic systems initially evolved by reducing entropy, and subsequently devised the means for conveying that knowledge from generation to generation through reproduction, then the key to understanding this process is that by reducing entropy, biology defies the second law of thermodynamics—over time, differences in temperature, pressure, and chemical potential tend to even out in a physical system that is isolated from the outside world. As a result, the cost shift in the bioenergetic expenditure toward the reproductive strategy has to ultimately be counterbalanced by a loss of bioenergetics in late life, since longevity is genetically determined (Hayflick 2007), leading to loss of cell–cell communication, culminating in death. But it should be emphasized that the biologic advantage of this strategy is that the organism genetically accumulates information from previous generations, while communicating genetic information to the next generation through reproduction.

The key is to follow the flow of biologic information through the communicative processes.

SOMEWHERE BETWEEN THE GENE AND THE PHENOTYPE LIES THE PROCESS OF EVOLUTION

It has been challenging to determine just what process lies between the gene and the phenotype. In the past, that mechanistic blackbox has been bridged by the use of such metaphors as natural selection, survival of the fittest, selection pressure, genetic assimilation, and adaptation. But these catch phrases do not reveal how such mechanisms work. In the era of genomics, we must determine the nature of such mechanisms for a number of reasons: (1) to determine the legitimacy of evolution, (2) to provide a predictive model for biology and medicine, and (3) to discover our origins in a rapidly changing world of global warming, stem cells, and interstellar space travel in search of new Earthlike planets.

In support of these pursuits, we have constructed a cellular–molecular context for evolution that, for the first time, provides a working model for the continuum of biology throughout the lifecycle. For example, it accounts for micro- and macroevolution, nurture and nature, contingency and emergence, and even the apparent duality of gradualism and punctuated equilibrium. The mere fact that the model can mechanistically account for such a broad range of evolutionary characteristics qualifies it for further consideration.

It is easier to describe biological evolution than it is to provide the scientific basis for its existence and relevance. Despite the academic revolution caused by merging developmental biology and evolution, or evo-devo, little actual hypothesis testing experimentation has been done to determine the validity of evolution theory. This has allowed advocates of intelligent design to gain an intellectual and emotional foothold in society, particularly in the United States, where a large number of people polled perennially reject evolutionary biology. Intelligent design is misguided and limiting because a belief cannot be falsified, nor can it lead to new thought. If we could devise experiments to test evolution theory, this would not only dampen the voices of the naysayers but would also widen the scope of scientific inquiry in the postgenomic era.

Most of the evolutionary biology literature is represented by either pure theory, such as (1) developmental systems theory, which is a collection of models of biological development and evolution that argue that the emphasis that modern evolutionary synthesis places on genes and natural selection as the explanation of living structures and processes is inadequate, or (2) by descriptive biology, including molecular mechanisms, but without their evolutionary origins. That is because we have not yet devised an effective way of thinking about evolution in real time. The core problem seems to be that evolution is conventionally considered to proceed from the present to the past, from right to left, whereas the experimental

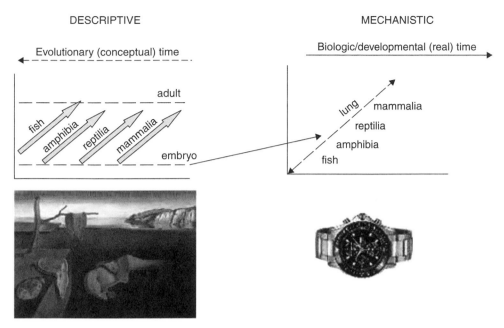

DESCRIPTIVE   MECHANISTIC

Figure 7.1. Thinking about evolution in Real Time. In order to make this conceptual transition to a paradigm in which evolutionary mechanisms can be tested, we must first focus on the embryonic stage within each phylum, and then begin to compare the cellular/molecular processes that give rise to structures and functions. (See insert for color representation.)

reality proceeds from the present to the future, from left to right. We suggest the following solution to this conceptual problem (Fig 7.1).

The classic representation for vertebrate evolution is as phyla, from fish to mammals—and, of course, we think of the adults in each phylum as being representatives of each group, yet evolution encompasses the entire lifecycle, from embryo to adult, including reproduction, ultimately returning to the unicellular state (the zygote). The description of phyletic evolution on the left of the schematic in Figure 7.1 allows us to organize the known biology in a first approximation of evolution, which does not lend itself to experimentation because it is based on chronological time and phenotypic homology, whereas experimentation must occur in real time. In order to make this conceptual transition to a paradigm in which evolutionary mechanisms can be tested, we must first focus on the embryonic stage within each phylum, comparing the cellular and molecular processes that give rise to structures and functions.

We use the lung as an example because we have begun doing such experimental evolution in our laboratory. We have found that the basic mechanism of lung morphogenesis changes progressively from fish to humans—parathyroid hormone–related protein (PTHrP) signaling intensity is amplified from the swim bladder of fish to the lungs of frogs, alligators, birds, and humans. The stepwise increase in

PTHrP signaling increases surfactant synthesis, which allows for the adaptive increase in alveolar surface area/blood volume ratio that facilitates the increase in gas exchange during vertebrate evolution, as first described by Clements et al. This phenomenon has been well documented in a series of publications by Daniels and Orgeig (2003).

A FUNCTIONAL GENOMIC APPROACH TO EVOLUTION AS AN EXAMPLE OF TERMINAL ADDITION

As mentioned in the previous chapter, Horowitz formulated a similar approach to the evolution of biochemical pathways by assuming a retrograde mode of evolution. That approach *describes* the functional phenotype for the evolution of a biosynthetic pathway, like the pathway labeled "Genetic" depicted at the bottom of the schematic in Figure 7.2 [labeled (c)]. In contrast to that, the cellular–molecular paracrine mechanism depicted as cell–cell interactions [labeled (a), steps 1–4], underpinned by a series of ligand–receptor interactions [labeled (b), steps 1–4] that evolved in response to a series of external (atmospheric oxygen, stretch) and internal (metabolic demand, tissue oxygenation, alveolar surface tension, blood pressure) selection pressures would have caused the evolution the homeostatic mechanisms that determined those biosynthetic pathways from phenotypes to genes. Selection pressure for such ligand–receptor-mediated gene regulatory networks would have generated both evolutionary stability and novelty through such well-known mechanisms as gene duplication, gene mutation, redundancy, alternative pathways, compensatory mechanisms, and positive and balancing selection pressures. Such phenotypic changes were in register with reproductive success, depicted as a progression of blue interlocking arrows. The genetic modifications were manifested by the structural and functional changes in the blood–gas barrier, primarily by the thinning of the blood–gas barrier in conjunction with adaptive phylogenetic changes in the composition of the surfactant, as described by Daniels and Orgeig (2003). The reverse engineering of these phenotypic changes in the blood–gas barrier form the basis for our molecular genetic approach to lung evolution. It should be noted that this approach provides an *emergent and contingent* mechanism for evolution for the first time.

More importantly, this model of cellular-molecular evolution predicts the evolution of other physiologic mechanisms by integrating reproduction into the selection pressure process—specifically, at each proximal step in the retrograde evolution of the surfactant, its physiologic roles, both in the newly evolved step and in other functionally interrelated aspects of integrated physiology, would have been constrained by the immediate and related mechanisms that prepare the embryo for its homeostatic adaptation to extrauterine life. For example, the relationship between stretch-regulated PTHrP signaling and surfactant production interrelates functionally (and genomically) with its complementary roles in bone

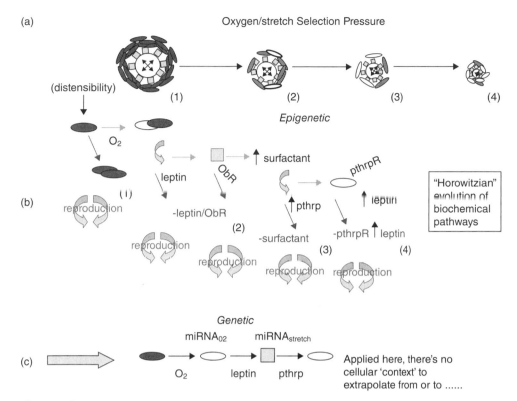

Figure 7.2. Evolutionary–developmental origins of the lung. At the top (a) is the cellular mechanism of alveolar development. In the middle (b) is an epigenetic paracrine cellular mechanism for alveolar evolution, reinforced by reproductive success (blue arrows forming cycle), showing how oxygen and stretch have driven a series of cellular interactions for lung evolution from the advent of the lipofibroblast to leptin production, causing increased surfactant production, which allowed facilitated the decreased alveolar diameter, increasing the surface area/volume ratio and increased gas exchange. At the bottom (c), a descriptive genetic pathway for the same process is shown. (See insert for color representation.)

development, skin maturation, the birth process, and glomerular physiology. Taking this one step further (backward), this same cellular–molecular processing from proximate to derived physiologic characteristics has been canonically repeated, perhaps all the way back to its unicellular origins—the cell-molecular mechanism of lung evolution based on the evolution of the surfactant dovetails with fundamental mechanisms of membrane evolution put forward by Konrad Bloch and by Cavalier-Smith (as discussed in Chapters 1 and 6). Bloch demonstrated that cholesterol evolved in response to the appearance of oxygen in the atmosphere, speculating that its biologic advantage was due to the reduced fluidity, or increased microviscosity resulting from the addition of cholesterol to the cell membrane phospholipid bilayer. The discovery of hopanoid triterpene derivatives in some prokaryotes, and in the form of "molecular fossils" of ancient times, has led to the suggestion (Rohmer 2008) that these relatively rigid, anaerobically evolved, amphiphilic molecules play a membrane reinforcement role in some prokaryotes similar to that played by aerobically evolved sterols such as cholesterol in eukaryotes. Bloom and Mouritsen hypothesized that the biosynthesis of cholesterol in the newly established aerobic atmosphere alleviated this constraint on the evolution of eukaryotes. On the basis of the observation by Cavalier-Smith that "there are twenty-two characters universally present in eukaryotes and universally absent from prokaryotes, presenting a detailed argument that, of these, the advent of exocytosis (and endocytosis) was the most likely to have provided the driving force for the evolution of eukaryotes into their present form," Bloom and Mouritsen (Miao et al. 2002) subsequently identified a possible constraint for cytosis, which was relieved by the advent of cholesterol. Therefore, there is a cell–molecule continuum from the evolution of cholesterol for the compliance of the plasma membrane of unicellular eukaryotes, to endocytosis/exocytosis, to the compliance of the evolving alveolar wall resulting from the secretion of cholesterol as a lubricant, to lung surfactant reducing surface activity.

This interrelationship between cholesterol, membrane thickness, and cytosis will appear again later in this book in considering the evolution of the surfactant, the increased oxygenation of the lung, and feeding.

SEEKING DEEP HOMOLOGIES IN LUNG EVOLUTION

The predictive value of PTHrP signaling in understanding vertebrate lung evolution has encouraged us to delve further into the role of this mechanism in alveolar evolution. Bearing in mind that during vertebrate lung evolution the alveoli have decreased in diameter in association with thinning of the alveolar wall, the natural process that transpires during lung development, we hypothesized that PTHrP would cause the thinning by stimulating mesenchymal apoptosis, or programmed cell death. Treatment of fetal rat lung fibroblasts with PTHrP did, indeed, inhibit expression of *bcl-2* and stimulate the expression of *bax*, which are the molecular characteristics of apoptosis. The observed effects of PTHrP on surfactant synthesis and apoptosis encouraged us to delve even deeper into this integrated cellular–molecular mechanism of alveolar evolution. In the mammalian lung, epithelial PTHrP signals to the lipofibroblast to stimulate the production of leptin, a hormone product of adipocytes in the periphery, and lipofibroblasts in the alveolar wall. The leptin then acts back on the alveolar epithelium to stimulate surfactant production, maintaining alveolar homeostasis as the alveolar wall expands and contracts. In an ongoing effort to understand the evolution of the vertebrate lung, we noted experiments showing that leptin stimulates limb development in frog tadpoles, mechanistically linking metabolism and locomotion, two of the three components of vertebrate evolution; the third component was respiration. Therefore, in an attempt to determine whether leptin would also affect lung development in amphib-

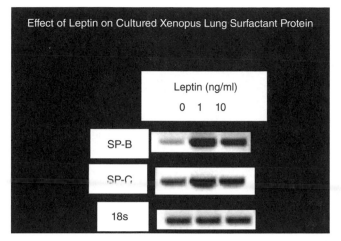

Figure 7.3. Effect of leptin on frog lung surfactant protein. Stage 55–57 tadpole lungs were treated with frog leptin (1 or 10 ng/mL) for 48 h in culture. Leptin treatment increased the expression of surfactant proteins B and C mRNA.

ians, we treated *Xenopus* lung tissue with frog leptin and examined its effects on the structure and function of the developing tadpole lung. To our amazement, we found that leptin had the same effects on structural development as in mammals, namely, the thinning of the blood–gas barrier, and increased alveolar surface area (Fig 7.3), in association with increased surfactant synthesis (Fig. 7.4). These effects were unexpected because the frog lung is not physiologically dependent on surfactant to lower surface tension, since the faveoli (the equivalent to alveoli in frogs) are so large in diameter, and are structurally surrounded by a muscular wall. Therefore, the results suggested that leptin primarily affected some other adaptive trait. Since surfactant proteins A and D are collectins, which function as antimicrobial peptides, we speculated that the primary selection pressure affected by the leptin was for host defense, which only secondarily affected surface-tension-lowering properties as an exaptation of this earlier adaptive trait.

The primary role of host defense in the evolution of the blood–gas barrier (BGB) may also relate to the commonalities between alveolar host defense and homeostasis. For example, we have observed that lipopolysaccharide (LPS), which constitutes the cell walls of Gram-negative bacteria, stimulates surfactant synthesis by alveolar type II cells, as does leptin, a member of the interleukin 6 (IL6) superfamily of cytokines. LPS stimulates toll-like receptor 4, triggering a host defense cascade characteristic of innate immunity. Evolutionarily, this may have been an early adaptation for the protection of the evolving lung as an outpouching of the gut, which would have allowed bacteria to infiltrate this structure and potentiate infection, particularly as the surface area of the lung later increased under selection pressure for increased gas exchange. We have hypothesized that leptin evolved as a product of the lipofibroblast, which, in turn, evolved to protect

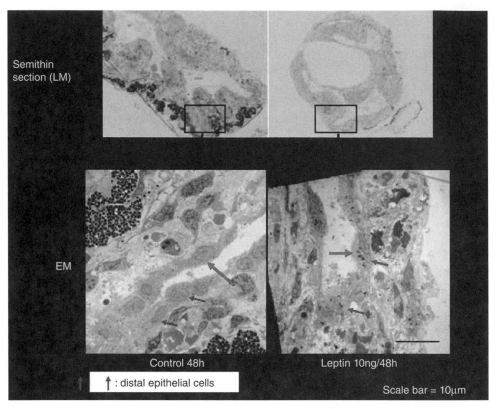

Figure 7.4. Effect of leptin on tadpole lung development. Stage 55–57 tadpole lungs were treated with frog leptin (10 ng/mL) for 48 h in culture. Light microscopic photomicrograph (LM, top, 40×) reflects increased faveolar size. Electron micrograph (EM, below) indicates thinning of epithelium (indicated by arrows) and more prominent lamellar bodies in the distal airways (arrows) in response to letptin treatment. (See insert for color representation.)

the alveolar wall against the increasing oxygen tension in the atmosphere in the Phanerozoic period. Leptin signals to the epithelial type II cell to produce surfactant, which acts to both protect against bacterial infection, mimicking the LPS mechanism, and increase alveolar compliance by stimulating surface-tension-reducing activity. In short, these mechanisms may have coevolved for these dually adaptive, physiologically synergistic purposes. Moreover, leptin stimulates epithelial cell proliferation, so it may also have positively selected for the evolutionary expansion of the blood–gas barrier (BGB).

The expansion of the BGB would have required increased membrane formation, including the generation of sphingolipids and cholesterol. The role of cholesterol in this evolutionary process is of particular interest because it has been implicated in a variety of processes, ranging from oxygen sensing, to cell growth and differentiation, adaptation to extracellular stress, lipid homeostasis, and finally, as antimicrobials—bacterial pore-forming toxins activate sterol regulatory element–binding protein 2 (SREBP2), inducing genes involved in lipid synthesis. This property of cholesterol is characterized in the literature as counterintuitive, and yet from an evolutionary perspective, this may be the primordial selection pressure, considering the importance of barrier function for survival. Even more importantly, this molecular functional linkage between cholesterol and host defense relates back to the primordial effects of cholesterol on the plasma membrane, endocytosis, cell–cell signaling (see Chapter 1), hormonal regulation, and the origins of complex physiology (see Chapter 6).

Another aspect of alveolar evolution has recently been highlighted by Maina and West (2005), who have speculated that the synthesis of alveolar type IV collagen must have increased during the evolution of the lung in order to compensate structurally for the progressive decrease in alveolar diameter. This phenotypic trait may have derived from the cholesterol mechanism, which increases plasma membrane stability by forming lipid rafts. Since the physiologic mechanism of PTHrP action on the developing alveolus is mediated by the upregulation of leptin expression by the mesoderm, which then acts in a retrograde paracrine fashion to stimulate alveolar epithelial differentiation, we hypothesized that leptin would stimulate type IV collagen synthesis. Leptin treatment of fetal rat lung alveolar type II cells stimulated type IV collagen synthesis, which is consistent with this working hypothesis. We observed a specific effect of leptin on type IV collagens 1α and 3α, suggesting that leptin promotes type IV collagen synthesis as an evolved trait for a thinner and stronger BGB. Another possibility is that leptin stimulates the epidermal growth factor receptor (EGFR), leading to upregulation of type IV collagen. The latter mechanism would be of more interest from an evolutionary mechanistic perspective because it would lead to other molecular pathways, in the lungs and other organs, that would interrelate lung evolution and lung pathophysiology. Moreover, the preadaptation for alveolar wall stability by cholesterol may have facilitated the EGFR type IV collagen mechanism since lipid rafts promote receptor density and signaling.

SYSTEMS BIOLOGY BASED ON CELL–CELL COMMUNICATION

In this book, we have formally proposed using a comparative, functional genomic approach to solve for the evolution of physiologic traits, which engenders development, homeostasis, and regeneration as a set of parallel lines that can be mathematically analyzed as a family of simultaneous equations (see Chapter 3). This perspective provides an *empirically refutable* way of systematically integrating such information in its most robust form to retrace its evolutionary origins.

Furthermore, we have demonstrated the utility of a middle–out approach for lung biology both as a basic tool for identifying novel functional genes and to target genes for the treatment of lung disease. For example, we have discovered that leptin is the intermediate for PTHrP signaling to adepithelial lipofibroblasts to stimulate surfactant. Leptin is a pleiotropic metabolic hormone that coordinates the development and evolution of locomotion and metabolism, since it stimulates limb development in frog tadpoles. We have now shown that it also stimulates tadpole lung development, providing an integrated model for both the ontogeny and phylogeny of these complex physiologic traits. Furthermore, breakdown in the PTHrP leptin signaling pathway in response to a wide variety of pathophysiologic agents (overdistension, hyperoxia, infection, nicotine) causes lung fibrosis, providing novel targets for the diagnosis, prevention, and treatment of chronic lung disease.

This cellular–molecular model of lung physiology is highly robust, possessing predictive power to extrapolate from the organ of gas exchange to the evolution of whole-animal physiology when similar mechanisms of selection pressure for cell–cell communication are applied (see Fig. 7.5). For example, we have found

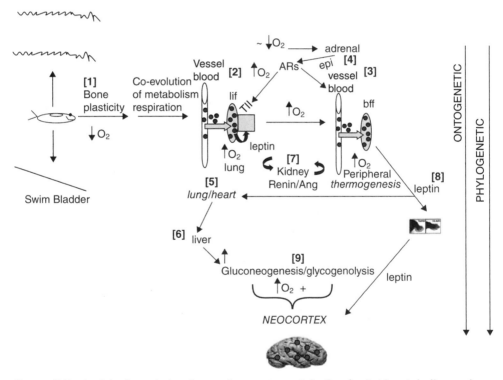

Figure 7.5. Physiologic evolution from cells to systems. Selection for lipid metabolism and respiration [1] gave rise to the lung; oxygen induced the lipofibroblast (lif) in the lung [2] and in the peripheral fat cells [3], putting selection pressure on lung/adrenal β-adrenergic signaling [4] to stabilize blood pressure (lung, periphery); coevolution of the lung and heart [5] induced the liver [6]; kidney rennin/angiotensin [7] provided further blood pressure stability. Leptin production by fat cells [8] promoted limb and lung development; increased oxygenation and glucose promoted neocortical development [9]. (See insert for color representation.)

that leptin increases the surface area of the developing frog lung in tandem with stimulation of surfactant production—the effect of leptin on both the antiatelectatic surface-tension-reducing mechanism, and on surfactant protein A, an antimicrobial peptide, points to the evolution of barrier function common to a variety of tissues and organs, such as gut, kidney, skin, and brain. Viewing these 'molecular phenotypes' across ontogenetic and phylogenetic space and time will lead to our understanding of the evolution of other such evolved physiologic interrelationships, ultimately generating a new paradigm for evolutionary biology and physiology (see Fig. 7.5).

Using these functional genomic precepts, we have generated a cellular–molecular model for vertebrate physiologic evolution, beginning with the coevolution of fish adaptation for feeding and buoyancy in adaptation to gravity as a preadaptation for PTHrP-regulated surfactant production, given that PTHrP is a gravisensor (step 1), and the swim bladder requires the surfactant, which is predominantly cholesterol, as a lubricant to inflate and deflate efficiently in adaptation to buoyancy (i.e., gravity) (step 2). The progressive metabolic demand for oxygen during vertebrate evolution, in combination with the large fluctuations in atmospheric oxygen during the Phanerozoic era (the last ~500 million years) created selection pressure for the increase in lung surface area, which was mediated by the increased efficiency for surfactant production, facilitating the increased surface area/blood volume ratio. It should also be borne in mind that the atmospheric temperature was also rising during this period of rising oxygen tension as a result of increased carbon dioxide in the atmosphere. This is consistent with the Romer Hypothesis that vertebrates were forced out of the water onto land because of the desiccation of bodies of water, which may have been caused by the rise in atmospheric temperature.

It has long been held that the transition from fish to tetrapods was due to the movement of vertebrates from water to land, and that fins progressively changed to limbs to accommodate terrestrial life. This concept was first formally proposed by the American geologist Joseph Barrell in 1916 (Barrell 1918), and later by the American paleontologist Alfred Sherwood Romer (Romer 1967). Both men surmised that fish were forced out of water onto land as Earth's climate became drier, beginning about 370 million years ago. More recently, Berner (1999) has provided empiric evidence for the episodic rises and falls in oxygen and carbon dioxide over the last ~500 million years during the Phanerozoic eon. As bodies of water dried up, fish had to develop a means of surviving on land, developing anatomical and physiologic characteristics in the process. The rising atmospheric temperature independently generated positive selection for respiration since lung surfactant is three fold more surface-active at 37°C than at 25°C.

Both oxygenation and surfactant production were facilitated by the timely emergence of the lipofibroblast, which evolved in response to the overall rise in atmospheric oxygen tension (Csete et al. 2001), protecting the alveoli against oxygen-free radicals, and stimulating surfactant production by the alveolar epithelium, rendering the alveolus more distensible, or compliant. The resulting rise in systemic oxygenation may, in turn, have caused the evolution of peripheral adipocytes (step 3), leading to endothermy as a systemic physiologic mechanism for increased compliance of the alveolar wall resulting from the adaptive effect of increased body temperature on surfactant activity mentioned above (i.e., step 3 synergizing step 2).

In support of the integration of whole-animal physiology by evolving fat cells, both the fat body of fruitflies and mammalian fat cells express the gene target of rapamycin (TOR), a molecular signaling pathway that monitors and responds to metabolic conditions. In response, fruitfly fat bodies produce a substance called *acid-labile subunit*, and mammalian fat cells produce leptin, both of which modulate insulin regulation of whole-body metabolism. The generation of such a systemic regulatory feedback mechanism may well have evolved through the positive selection pressure effect of fat cells for both respiration and endothermy. In turn, these molecular fat cell phenotypes may have generated positive selection pressure for neocortical brain evolution, since lipids are necessary for myelination and neuronal signal transduction, facilitating integrated control of respiration, body temperature, and other evolving physiologic functions. Mammals and birds further evolved novel physiologic characteristics that were dependent on such neuronal evolution.

VERTICAL INTEGRATION OF LEPTIN SIGNALING, HUMAN EVOLUTION, AND THE TROJAN HORSE EFFECT

Leptin affects the growth of a wide variety (Fig. 7.6) of tissues and organs, ranging from blood vessels, to bone, and neurons in both the peripheral and central nervous systems. The robust, predictive nature of the cellular–molecular approach to understanding integrated physiology is underscored by the commonality between fat metabolism in both the peripheral and central nervous systems, where lipid metabolism has been positively selected for the evolution of the brain neocortex. At the cellular–molecular level, for example, neuregulin is a soluble paracrine product of nerve axons that regulates lipid incorporation into myelin by Schwann cells, promoting nerve activity. Neuregulin also controls lung surfactant lipid synthesis through paracrine signals mediating the interaction between connective tissue and epithelial cells. Experimentally, treating leptin-deficient mice with leptin stimulates respiration, and feeding activity in metamorphosing frogs, both of which are controlled by the central nervous system. Furthermore, leptin specifically affects the neurons that innervate the hypothalamus, programming γ-1-syntrophin (SNTG1) signaling through diacylglycerol kinase-ζ, which regulates serotonin (Fig. 7.6). Consistent with the positive selection pressure for leptin signaling, the leptin receptor, its signaling pathway, and SNTG1 have been found to be rapidly evolving in human populations (Voight et al. 2006). Thus leptin determines the mechanisms of cell–cell signaling at multiple levels, integrating

Vertical Integration of Molecular-Cellular and Evolutionary Change

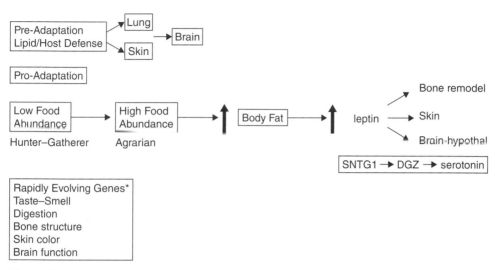

Figure 7.6. From lipids to brain evolution. This model of human evolution starts from the premise that cholesterol was key to vertebrate evolution, and traces the influence of lipid metabolism from host defense to myelination and hypothalamic evolution for central nervous system control of physiology. This model explains the interrelationships between a variety of genes found to be rapidly evolving in human populations by Voight et al. (2006).

physiology from cells to tissues and organs, and from the peripheral to the central nervous systems, all the way to hypothalamic serotonin, which determines animal behavior on the basis of the pleasure principle.

LEPTIN AND HUMAN EVOLUTION: FOOD FOR THOUGHT

These leptin-based molecular phenotypes projecting all the way back to unicellular organisms may ultimately have advanced human evolution via a fundamental change in the human diet. At the whole-organism level, as humans evolved from hunter-gatherers to farmer some approximately 7000–16,000 years ago, the sporadic nature of food resources became more constant. This changed human metabolism, resulting in sustained, elevated production of circulating leptin produced by increased body fat deposits. That would have facilitated the more recent historic increases in both overall body size and neocortical development that accompanied the advent of hydraulic societies in the Far East, Middle East, and Central America. The advent of the first and second industrial ages, followed by the modern and postindustrial eras, freed humans from much of their need for physical exertion for food gathering, mobility, warmth, and other basic necessities. As a result, the selection advantage for physical traits was supplanted by mental exercise, placing increasing selection pressure on the neocortex. Therefore, higher circulating levels of glucose are advantageous, since the brain is a glucose sink. Of course, too much of a good thing may lead to type II diabetes, so the elevation of blood glucose must be temporized by insulin regulation.

When these concepts of human physiologic and social evolution are coupled with emerging knowledge of brain plasticity and stem cells, it is no wonder that humans evolved so rapidly over the intervening period. As proof of principle, a study of more recent rapidly evolving human genes (Voight et al. 2006) revealed, for example, that these genes included both the leptin receptor and the phosphatidylinositol pathway genes, which mediate leptin receptor signal transduction, corroborating the significance of leptin's role in human evolution. Positive selection for both the leptin receptor and the leptin receptor signaling mechanism epitomizes the concept of adaptive evolution mediated by cell–cell communication.

However, another gene that has been found to be rapidly evolving among humans is the cytochrome P450 gene CYP3A5, which we had noted in Chapter 5 (section on salinity and the vitamin D receptor), to be involved in the counterbalancing selection pressure for vitamin D metabolism. Recall that in that scenario for human migration, there was positive selection pressure for CYP3A5 as a means of inhibiting the effect of salt on blood pressure; that same gene has been shown to be closely associated with the occurrence of schizophrenia. Therefore, early in human evolution (Fig. 7.7) during the hunter-gatherer phase selection pressure for vitamin D mediated by skin pigmentation was benefitted by the CYP3A5 gene, whereas when humans began farming, and the selection pressure favored intelligence, CYP3A5 may have led to the increased incidence of schizophrenia, although only 1% of the population are afflicted, so perhaps it was worth the risk.

The perpetuation of such a maladaptive gene occurred because it was disguised as adaptive for the skin–vitamin D metabolic trait, acting like a Trojan horse, which allowed it to be carried along in the human gene pool. Phenotypic commonalities between skin, lung, and brain can be found at the molecular level. For example, there is overlap between the lung and skin as barriers, both of which have packaged lipids together with β-defensins for host defense. These selection pressures then telescope to neocortical myelination, as shown above. They also translate into disease. For example, patients with asthma often have the skin disease atopic dermatitis. These phenotypes have been mechanistically linked through a common molecular defect in β-defensin, which mediates innate host defense in both skin and lung.

In dogs, β-defensins have been shown to determine coat color, which serves a multitude of adaptive advantages, ranging from protective coloration to reproductive strategies. β-Defensin CD103 has also been shown to cause atopic dermatitis in dogs, and possibly asthma, since it is found in dog airway epithelial cells. We will discuss the implications of such a Trojan horse effect in Chapter 10.

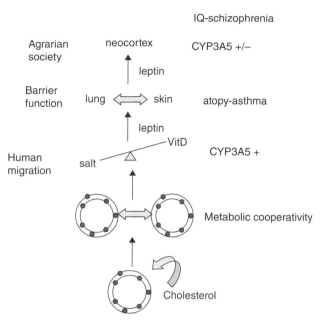

Figure 7.7. Trojan horse effect and human evolution. Over the course of vertebrate evolution, starting with the introduction of cholesterol in single-celled eukaryotes, early human migration was facilitated by CYP3A5, but later neocortical evolution driven by leptin introduced schizophrenia as a maladaptation through the Trojan horse effect. (See insert for color representation.)

In comparison with the lipid pathway for neocortical evolution, dogs and diving mammals developed different traits in adapting to the increased availability of food, provided by humans in the case of dogs, and by returning to water in the case of diving mammals. The differences in the arc of evolution in these species underscore the emergent and contingent nature of the evolutionary process. This may be due to the contingency of leptin signaling through, for example, the epidermal growth factor pathway for cerebral evolution in humans versus the insulin-like growth factor pathway for size in dogs. We do not know what molecular genetic characteristics have distinguished humans from other primates yet, but such differential gene expression may hold the answer.

Moreover, Kluger (Kozak et al. 2000) has shown that thermogenesis evolved to combat infection, a molecular phenotype that refers all the way back to the balancing selection for host defense between salt-sensitive antimicrobial peptides, vitamin D metabolism, and the subsequent evolution of leptin production by lipofibroblasts for lung development described in Chapter 5 (see Fig. 5.1 and text in section on the continuum from microevolution to macroevolution).

At a critical phase in lung evolution, the physiologic bottleneck may have been the sensitivity of the microvasculature to elevated systemic blood–pressure, and the thickness of the blood–gas barrier for gas exchange. Further evolution of the lung was then facilitated by the expression of type IV collagen, increasing the

tensile strength of the alveolar wall, which seems to have occurred phylogenetically sometime between the emergence of fish and amphibians, based on the
appearance of the Goodpasture epitope for type IV collagen during this transition
(see in-depth discussion in Chapter 6, section on possible a molecular evolutionary
link between the lung and kidney). This was followed by the selection pressure
for β-adrenergic receptors, which allowed for the independent regulation of blood
pressure in the lung and in the systemic circulation. Consideration of this particular
environmental constraint is particularly instructive because of the biologic effects
of alternating atmospheric hyperoxic and hypoxic conditions that have occurred
during the Phanerozoic eon. The successful metabolic adaptation to a hyperoxic
environment is manifested by an increase in the surface area for gas exchange
surface area. This was subsequently followed by a period of relative hypoxia,
which is the most potent physiologic β-adrenergic agonist (Fig. 7.5, step 4). Those
organisms that were able to survive these whipsawing physiologic effects of alternating hyperoxia and hypoxia by structurally and functionally adapting form the
basis for the intrinsic and extrinsic selection pressures for the phylogeny and
ontogeny of lung evolution depicted in Figure 5.1. It is at this phase in vertebrate
evolution that the glucocorticoid receptor is documented to have evolved from the
mineralocorticoid receptor, perhaps as a counterbalancing selection for the blood
pressure–elevating effect of mineralocorticoids. The emergence of the glucocorticoid mechanism may have been hindered by the presence of pentacyclic triterpenoids generated by rancidifying land vegetation. These compounds inhibit
11βHSD2, the enzymatic inactivator of cortisol's blood pressure–stimulating
activity, causing selection pressure for the tissue-specific expression of 11βHSD1,2,
permitting the local tissue activation and inactivation of cortisol. Accomodation
of blood pressure control may have been further facilitated by the evolution of the
rennin–angiotensin system of the kidney (Fig. 7.5, step 7). The adipocyte-derived
metabolic hormone leptin (Fig. 7.5, step 8) would have acted pleiotropically to
facilitate locomotion and respiration, representing the three elements of land vertebrate evolution, placing increasing selection pressure on the cardiac and pulmonary systems, which evolved in tandem with one another (Fig. 7.5, step 5) from
one-, to three-, and then four-chambered hearts. The increased heart complexity
was accompanied by increased heart size, causing the heart to precociously
impinge on the gut mesoderm, inducing progressively earlier liver differentiation
(Fig. 7.5, step 6), providing a controlled source of glucose from stored glycogen
under endocrine control (Fig. 7.5, step 9) for brain development in combination
with increased oxygenation. The evolution of the vertebrate brain would have
provided the obligate central control for the further integrated physiologic evolution (Fig. 7.5, steps 2–7).

It is important to point out that the determinants of the evolution of the lung
depicted in Figure 7.2, and in other tissues, organs, and systems depicted in Figure
7.5, are based on fundamental principles of physics—gravity, gas laws, and
thermodynamics—which Darwin called *natural selection*. Therefore, the mechanisms of micro- and macroevolution operate on the same principles.

In Chapter 8 we will discuss the possibility that the networks of genes derived from the algorithm for integrated physiology described in this chapter could be used to generate a periodic table for biology, linking specific phenotypes together in ways that would be counterintuitive in terms of conventional descriptive biology. Such an approach offers a dynamic and novel way of thinking about how the genomic elements of physiologic systems may have been recombined during evolution to create novelty based on cellular principles of phylogeny and development, rather than on relatively static descriptions of structure and function.

REFERENCES

Barrell J (1918), *The Evolution of the Earth and Its Inhabitants: A Series of Lectures Delivered before the Yale Chapter of the Sigma Xi during the Academic Year 1916–1917.* Yale University Press, New Haven, CT.

Berner RA (1999), Atmospheric oxygen over Phanerozoic time. *Proc. Natl. Acad. Sci. U.S.A.* 96(20):10955–10957.

Csete M, Walikonis J, Slawny N, Wei Y, Korsnes S, Doyle JC, Wold B (2001), Oxygen-mediated regulation of skeletal muscle satellite cell proliferation and adipogenesis in culture. *J. Cell Physiol.* 189(2):189–196.

Daniels CB, Orgeig S (2003), Pulmonary surfactant: the key to the evolution of air breathing. *News Physiol. Sci.* 18:151–157.

Hayflick L (2007), Biological aging is no longer an unsolved problem. *Ann. NY Acad. Sci.* 1100:1–13.

Kozak W, Kluger MJ, Tesfaigzi J, Kozak A, Mayfield KP, Wachulec M, Dokladny K (2000), Molecular mechanisms of fever and endogenous antipyresis. *Ann. NY Acad. Sci.* 917:121–134.

Maina JN, West JB (2005), Thin and strong! The bioengineering dilemma in the structural and functional design of the blood-gas barrier. *Physiol. Rev.* 85(3):811–844.

Miao L, Nielsen M, Thewalt J, Ipsen JH, Bloom M, Zuckermann MJ, Mouritsen OG (2002), From lanosterol to cholesterol: structural evolution and differential effects on lipid bilayers. *Biophys. J.* 82(3):1429–1444.

Rohmer M (2008), From molecular fossils of bacterial hopanoids to the formation of isoprene units: Discovery and elucidation of the methylerythritol phosphate pathway. *Lipids* 43(12):1095–1107.

Romer AS (1967), Major steps in vertebrate evolution. *Science* 158(809):1629–1637.

Voight BF, Kudaravalli S, Wen X, Pritchard JK (2006), A map of recent positive selection in the human genome. *PLoS Biol.* 4(3):446–458.

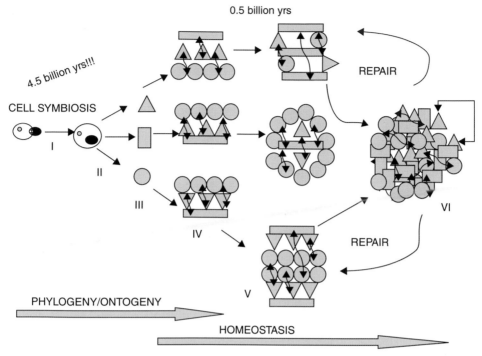

Figure 1.1. Cooperative cells as the origin of vertebrate evolution. (See text for full caption.)

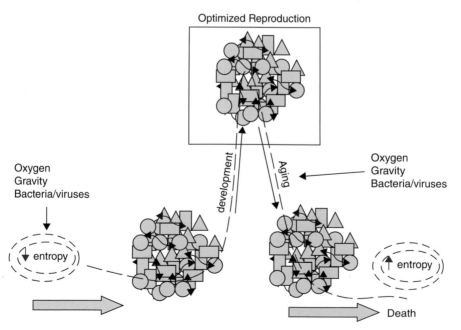

Figure 1.3. Evolutionary selection pressure, development, homeostasis, and aging. (See text for full caption.)

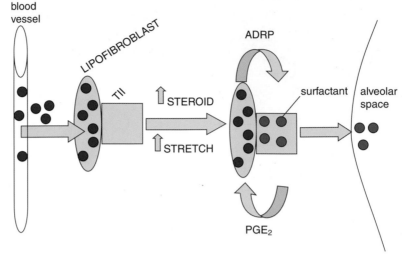

Overall Mechanism for Neutral Lipid Trafficking from Lipofibroblast to Type II Cell

Figure 2.1. Neutral lipid trafficking from lipofibroblast to type II cell. Lipofibroblasts actively take up and store neutral lipid from the circulation by expressing adipocyte differentiation–related protein (ADRP). Alveolar type II cells (TII) recruit these neutral lipids by secreting prostaglandin E_2. Hormones and mechanical stretching coordinately regulate these mechanisms to integrate distension of the alveoli during breathing with the production of lung surfactant. (From Torday et al. 1995.)

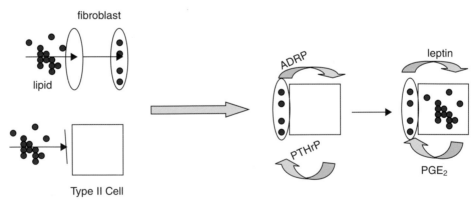

Figure 2.2. Experimental evidence for neutral lipid trafficking. Monolayer cultures of lung fibroblasts actively take up neutral lipid, but don't release them (top left) unless they are cocultured with type II cells. Type II cells cannot take up neutral lipid (lower left) unless they are cocultured with lung fibroblasts. Fibroblast uptake of neutral lipid is determined by PTHrP from the type II cell, which stimulates ADRP expression by the lung fibroblast (middle); stretching cocultured lung lipofibroblasts and type II cells increases surfactant synthesis by coordinately stimulating PGE_2 production by type II cells, causing release of neutral lipid by the lipofibrfoblasts, and leptin secretion by the lipofibroblasts, which stimulates surfactant phospholipid synthesis by the type II cells.

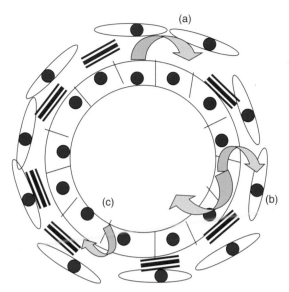

Figure 3.2. Effects of PTHrP on the evolving alveolus: (a) inhibition of fibroblast growth; (b) stimulation of surfactant; (c) stimulation of type IV collagen.

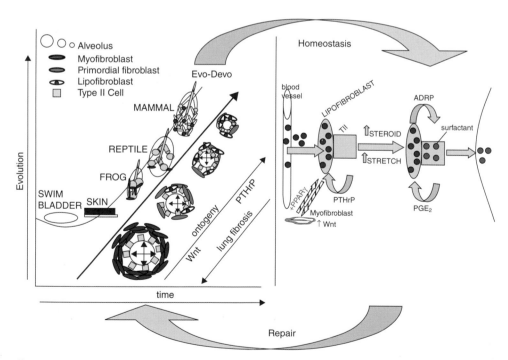

Figure 4.1. Lung biologic continuum from ontogeny–phylogeny to homeostasis and repair. (See text for full caption.)

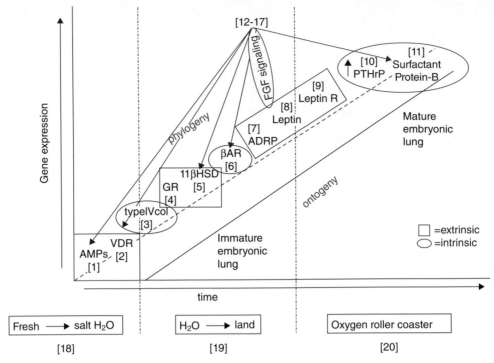

Figure 5.1. Alternating extrinsic and intrinsic selection pressures for the genes of lung phylogeny and ontogeny. The effects of the extrinsic factors (salinity, land nutrients, oxygen) on genes that determine the phylogeny and ontogeny of the mammalian lung alternate sequentially with the intrinsic genetic factors, highlighted by the circles and squares.

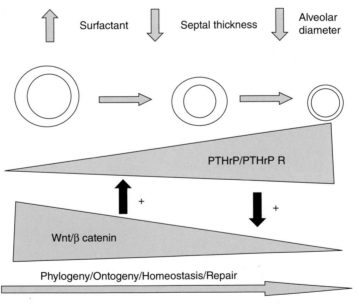

Figure 5.2. Progressive evolutionary decreases in alveolar septal thickness and diameter. Over the evolutionary history of the lung alveolus, its phylogeny and ontogeny have been shaped by selection pressure for increased surfactant production, mediated by Wnt/β-catenin and PTHrP/ PTHrP receptor paracrine signaling.

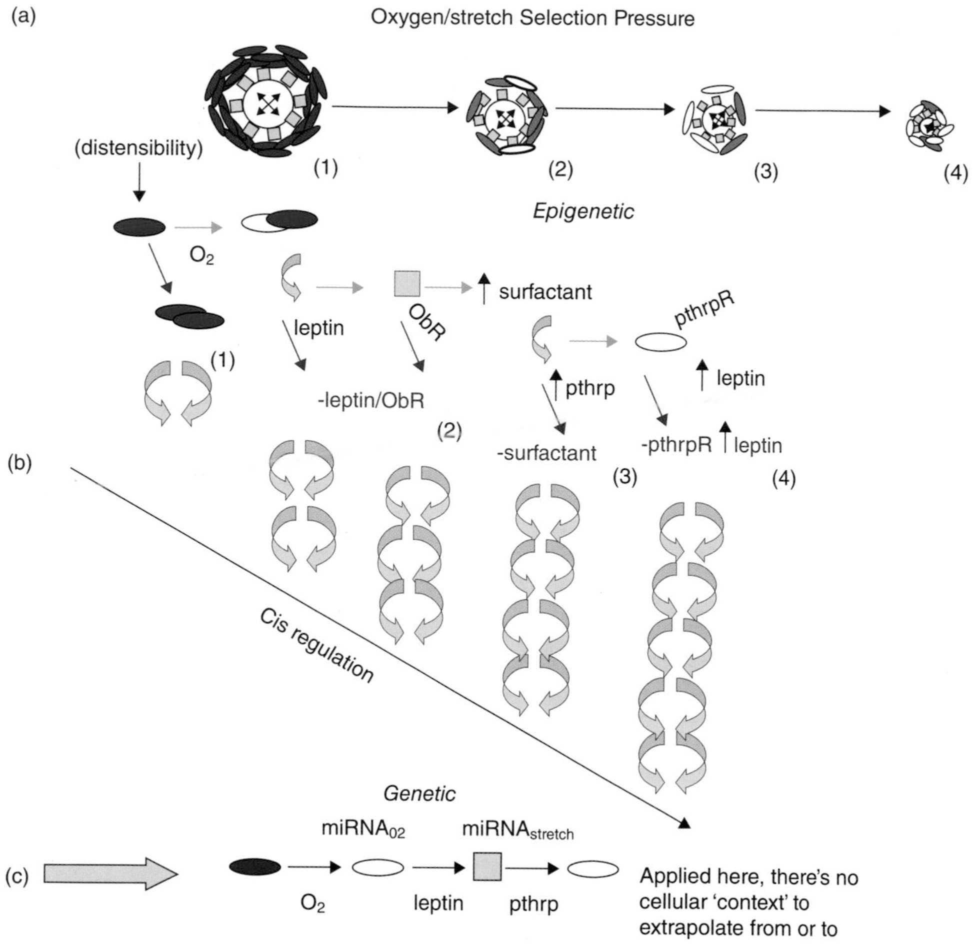

Figure 5.4. Evolutionary–developmental origins of the lung. The top portion (a) depicts the paracrine cellular mechanism of alveolar development, phylogeny, and evolution. The middle portion (b) depicts the cell–cell signaling mechanisms that have facilitated alveolar evolution, mediated by soluble factors such as leptin and PTHrP, and the *cis* regulatory elements that had to evolve. The cascade shows how oxygen and stretch may have driven a series of cellular interactions for lung evolution, from the advent of the lipofibroblast to leptin production, causing increased surfactant and increased distensibility, placing positive selection pressure on stretch regulation of PTHrP. This process ultimately allowed for the progressive decrease in alveolar diameter, increasing the the surface area/volume ratio and increased gas exchange. Portion (c) shows a traditionally descriptive genetic pathway for the same processes shown in (a) and (b).

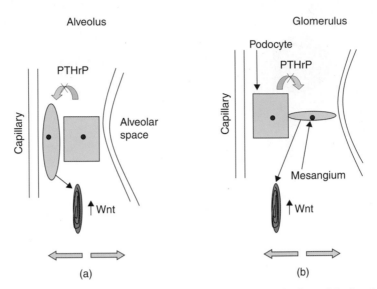

Figure 6.3. The alveolus and glomerulus are stretch sensors. In the lung (a), the alveolar epithelium (square) and fibroblast (oval) respond to the stretching of the alveolar wall by increasing surfactant production. In the kidney (b), the mesangium (oval) senses fluid pressure and regulates bloodflow in the glomeruli. In both cases, breakdown in cell–cell interactions causes these cells to become fibrotic (brown cell) due to upregulation of Wnt.

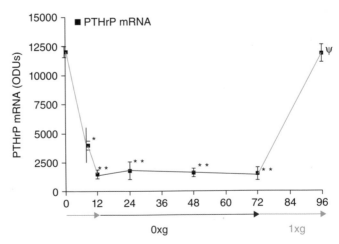

Figure 6.4. Human UMR 106 bone cells were maintained in a rotating wall vessel for ≤ 72 h. At the end of that time, the cells were returned to unit gravity (1 × g) for 24 h. Cells were analyzed for PTHrP mRNA expression using reverse transcriptase polymerase chain reaction (RT-PCR). [*Notation*: $N = 6$; *, $p < 0.00001$; **, $p < 0.000001$; Ψ, $p > 0.05$ vs. time = 0 by analysis of variance. The units on the y axis (ordinate) labeled ODU stand for optical density units.]

Evolutionary (conceptual) time

Biologic/developmental (real) time

adult

lung mammalia

reptilia

amphibia

fish

fish

amphibia

reptilia

mammalia

embryo

Figure 7.1. Thinking about evolution in Real Time. In order to make this conceptual transition to a paradigm in which evolutionary mechanisms can be tested, we must first focus on the embryonic stage within each phylum, and then begin to compare the cellular/molecular processes that give rise to structures and functions.

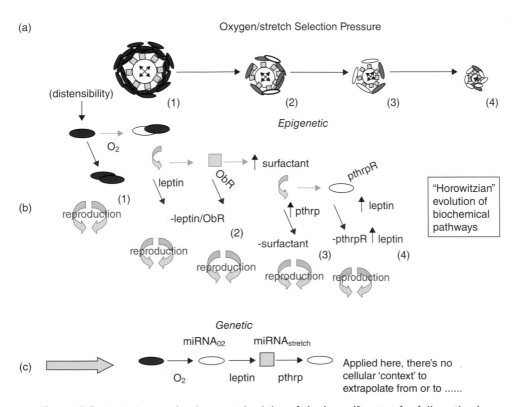

Figure 7.2. Evolutionary–developmental origins of the lung. (See text for full caption.)

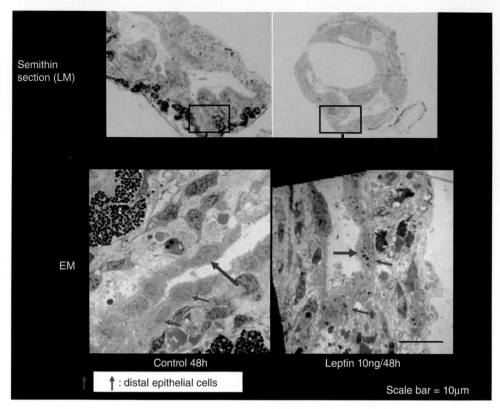

Figure 7.4. Effect of leptin on tadpole lung development. (See text for full caption.)

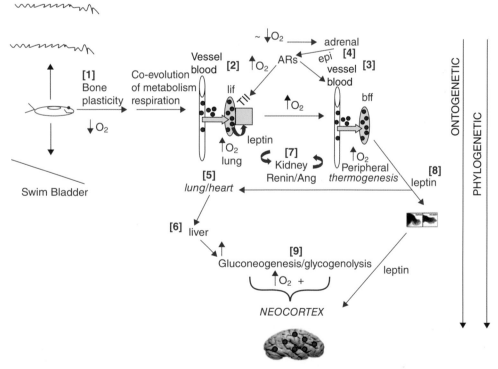

Figure 7.5. Physiologic evolution from cells to systems. (See text for full caption.)

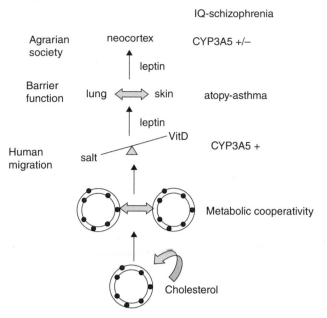

Figure 7.7. Trojan horse effect and human evolution. (See text for full caption.)

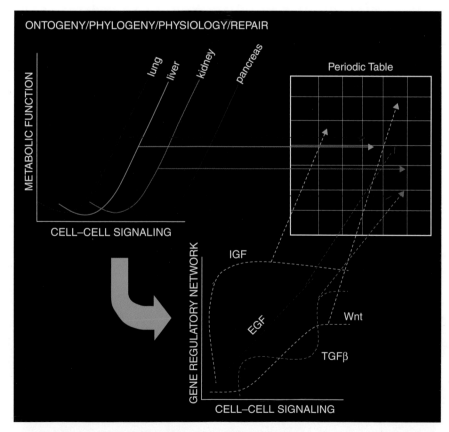

Figure 8.1. The biologic periodic table. In the upper left-hand corner, cell–cell signaling for the ontogeny, phylogeny, and repair of various organs is regressed against metabolic function. Below, the descriptive data are replaced by mediators of the signaling pathways. To the right, the periodic table of biology is depicted as a four-dimensional hologram.

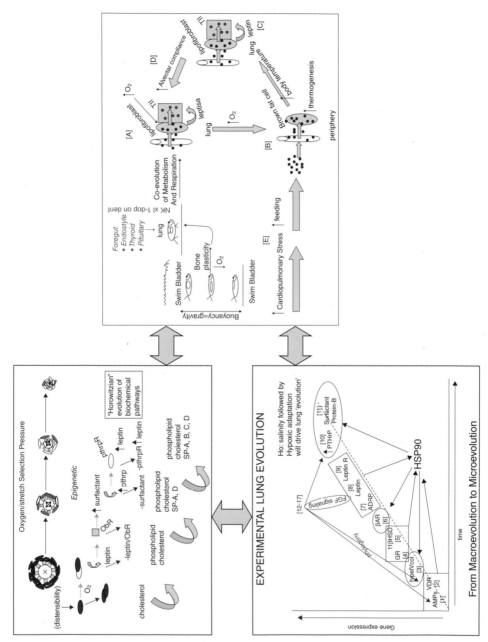

Figure 8.2. Merging schematics for cellular, genomic, and physiologic evolution. The schematics from earlier chapters are conceptually merged together.

How Genes Recombine to Generate Novel Phenotypes

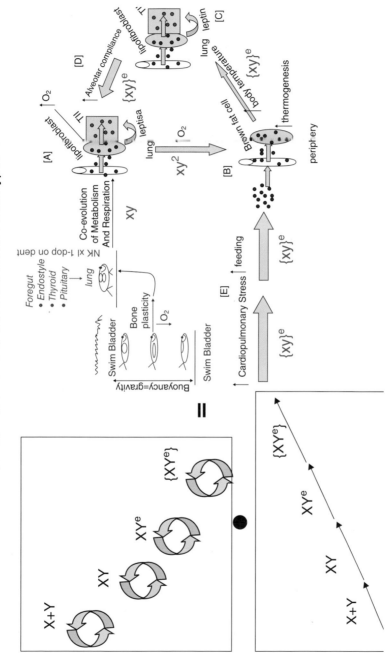

Figure 8.3. Assignment of algebraic expressions for the evolutionary integration of cellular, genomic, and physiologic evolution. In anticipation of mathematical expression of the evolution model, we have assigned algebraic characters for semiquantitation.

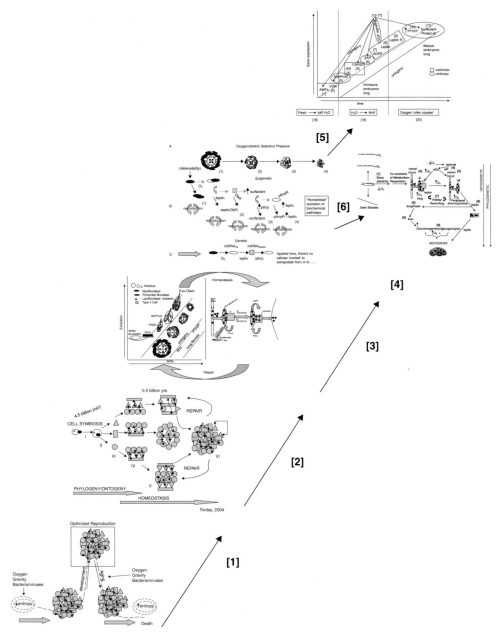

Figure 9.3. Integration of physiology. Continuum from [1] the advent of the cell, to [2] cell–cell signaling for multicellularity, [3] lung cell evolution based on surfactant production, [4] evolution of *cis* regulation, [5] intrinsic–extrinsic selection pressure, and [6] the lung as a cellular–molecular basis for physiologic evolution.

8

THE PERIODIC TABLE
OF BIOLOGY

THE PROSPECT OF A PERIODIC TABLE OF BIOLOGY

Humankind has long sought to gather scientific information and impart order to it, from Denis Diderot's *Encyclopedia* (1750), to Carl Linnaeus' binomial nomenclature for the classification of animals and plants, to today's numerous genomic, proteomic, and metabolomic annotations.

Several physicists in the nineteenth century had attempted to order the known chemical elements. Notably, Dmitri Mendeleev published his particular version

Evolutionary Biology, Cell–Cell Communication, and Complex Disease, First Edition.
John S. Torday and Virender K. Rehan.
© 2012 Wiley-Blackwell. Published 2012 by John Wiley & Sons, Inc.

of the Periodic Table and law in 1869. Its most striking features were that it predicted both how the elements would behave in chemical reactions and that there were missing elements.

In the same way that Mendeleev and other chemists and scientists have systematized the functional organization of the elements, biologists must systematize the burgeoning life science informatics databases. Currently, this is being done using descriptive, statistical analyses of high-throughput genomic data, which assumes that genes are expressed as a result of random occurrences. This is the legacy of the original interpretation of natural selection as a function of genetic mutation (microevolution) and population genetics (macroevolution). As mentioned previously, Albert Einstein famously stated in reference to the value of quantum mechanics and Heisenberg's uncertainty principle that "God does not play dice with the universe." The gene pools of contemporary species have not evolved by chance, but rather via the process of natural selection acting through a combination of internal and external selection pressures generating phenotypic diversity, as discussed in Chapter 7. Therefore, we should be thinking about how to exploit this process in order to discern gene selection patterns on the basis of functional and comparative genomics.

Our own research has approached lung development and homeostasis with respect to the intercellular signaling partners involved in lung alveolization. Disrupting these lung developmental and homeostatic signaling pathways results in chronic lung disease. Additionally, studies have shown that these same pathways have facilitated the evolution of the vertebrate gas exchange organ, beginning with the swim bladder of fish. Such commonalities indicate a continuum from ontogeny and phylogeny, to physiology and regeneration–repair that serves as a framework for vertically integrating cell–cell signaling across all of these processes. As a result, it provides a way of hierarchically organizing such data, leading to a predictive model of biology based on *a priori* knowledge, rather than on descriptive biology and statistics.

CELLULAR COOPERATION IS KEY

The cooperativity that underlies endosymbiosis in the rise of eukaryotes has evolved from a metabolic form, to a cellular form that has been recapitulated throughout the evolutionary history of multicellular organisms, as well as vertebrate phylogeny and ontogeny.

Take, for example, the epithelial–mesenchymal interactions that form the tissues and organs. Such interactions are necessary for the formation of the liver, as well as its homeostatic control of lipids, which traffic back and forth between the hepatic stellate cells and hepatocytes. Cell–cell interactions that control development and regulation of endocrine tissues such as the adrenals, gonads, prostate, and mammary gland can be viewed similarly.

In the lung development–homeostasis model that we have devised (see Chapter 2), lipid trafficking maintains the structural integrity of the alveoli. Surfactant, a lipid–protein complex, is produced by epithelial type II cells in the corners of the alveoli. As the lung expands and contracts in response to inspiration and expiration of air, physical distension of the alveoli regulates surfactant production and secretion. To maintain homeostasis, the connective tissue cells of the alveolar wall actively recruit lipids from the circulation, store them, and transport them to the epithelial type II cells in response to the physiologic demand for surfactant phospholipid synthesis (see Chapter 2).

This process is mediated by adipocyte differentiation–related protein (ADRP), which is under the control of the parathyroid hormone–related protein (PTHrP) signaling pathway. These functionally interrelated proteins are expressed compartmentally, as follows: PTHrP, surfactant, and the leptin receptor are expressed in the alveolar epithelium; the PTHrP receptor, ADRP, and leptin are expressed by lipofibroblasts. During lung development, these proteins appear sequentially in the epithelium and fibroblasts, beginning with epithelial PTHrP, followed by PTHrP receptor expression, ADRP, and leptin in the fibroblasts. The leptin receptor appears subsequently in the epithelial type II cells, mediating the stimulatory effect of leptin on lung surfactant production, which is necessary for the transition of the fetus from the amniotic sac to air breathing. This complementary spatiotemporal sequence of gene expression for ligands and their receptors on epithelial and mesenchymal cells appearing during development, or ontogeny, may also reflect their phylogeny since these genes appear in the same sequence in both processes. If so, the progression would represent the cellular–molecular evolution of the lung. Absence of any one of these regulatory proteins can cause respiratory distress syndrome due to surfactant deficiency, and death attributable to oxygen insufficiency if the infant is not placed on a mechanical ventilator. If the infant survives, it usually develops the chronic lung disease of prematurity referred to as *bronchopulmonary dysplasia* (BPD). A number of other pregnancy and postpregnancy factors can also disrupt the normal development of the signaling proteins that regulate lung surfactant. For instance, before birth, maternal smoking can interfere with alveolar development through the action of nicotine from the maternal circulation, which crosses the placenta to the fetus, specifically interfering with the signaling mechanisms that determine the development of the lung epithelial cells and fibroblasts. Infection of the fetus has similar effects on lung development. After birth, exposure of the preterm infant to mechanical respiration can also disrupt these signaling mechanisms. For example, increased distension of the alveoli, or *barotrauma*, can inhibit PTHrP signaling to its receptor, inhibiting the development of the lipofibroblasts, as well as the production of leptin needed to stimulate surfactant production by the epithelial type II cells. Additionally, the administration of oxygen, which is necessary for survival of preterm infants, can also inhibit the cell–cell signaling mechanism for surfactant production. Interrupting the cellular crosstalk causes epithelial and mesodermal cells to readapt as the pathologic process of BPD. A similar chain of events occurs in all structures that depend on such developmental cell–cell interactions. The recognition that ontogeny, phylogeny, and pathophysiology are on a genetic signaling continuum suggests that such motifs represent general patterns that could serve as guidelines for the assembly of a biologic periodic table.

ELEMENTAL BIOLOGY

The genius of the Periodic Table of Elements lies in its hierarchical organization, with atomic number as its organizing principle. A comparable approach to the creation of a biologic periodic table would need to be based on specific functional principles of homeostasis, linked together mechanistically through the genes that determine such processes. For example, our laboratory has taken a developmental approach to understanding the origins, homeostatic control, and pathophysiology of the lung, based on the vertically integrated effects of PTHrP.

During development, PTHrP signaling to its receptor is stimulated by fluid distension of the lung in the womb, and its signal is amplified by stimulating cyclic AMP-dependent protein kinase A (PKA) through multiple pathways for differentiation of the lung fibroblasts. The fibroblasts then produce factors that coordinate the growth and differentiation of the neighboring epithelial and vascular compartments, culminating in normal physiologic function, or homeostasis. Disruption of this hierarchy causes failure of the same cellular crosstalk mechanisms that mediate development and homeostasis, indicating a mechanistic continuum from disease to development. The inference is that by targeting the specific cellular–molecular physiologic elements of normal lung development, BPD can be either prevented or treated effectively.

PTHrP AS AN ARCHETYPE

The recognition that development, phylogeny, homeostasis, and pathophysiology constitute a continuum of functionally interrelated genes, signaling through specific cellular ligand–receptor pathways, provides the opportunity to consider how these cellular and molecular motifs have been retained through convergent evolution. We have hypothesized that the progressive complexity of the gas exchange unit has resulted from the phylogenetic amplification of the PTHrP signaling pathway (see Chapter 4). PTHrP and its receptor are expressed as early in vertebrate phylogeny as the swim bladder of fish, as is the expression of surfactant. However, in fish the swim bladder functions for buoyancy in adaptation for feeding, not to provide oxygen for metabolism. The swim bladder doesn't require the surfactant to prevent atelectasis, yet it expresses both PTHrP and surfactant as housekeeping genes. It is preadapted, and the regulatory genes function as housekeeping genes at this stage in the evolution of the lung.

During lung evolution from the fish swim bladder, the structure–function inter-relationship between gas exchange surface area and surfactant production has been amplified. We have hypothesized that both of these properties of the lung are mechanistically linked through stretch regulation of PTHrP signaling. Cyclic stretch coordinately enhances the expression of both epithelial PTHrP and the PTHrP receptor on the neighboring mesodermal fibroblasts.

As the metabolic demand on vertebrates drove their evolution, the lung surfac-tant system became progressively more efficient through selection pressure for the PTHrP-mediated effect on lung surfactant production, as follows: PTHrP ampli-fication of surfactant, along with PTHrP inhibition of fibroblast growth, may have promoted the thinning of the alveolar septa, allowing for the progressive phylo-genetic and developmental decreases in alveolar size. Either the resulting progres-sive increase in surface tension [by the law of Laplace, which states that surface tension is inversely related to the diameter of a sphere (i.e., alveolus)] had to be compensated for through balancing selection for the stretch-regulated surfactant mechanism, or the organism became extinct.

Gene knockouts yield evidence that a variety of tissues use such signaling strategies between cells of different embryonic germline origins for varied func-tions. For example, PTHrP deletion gives rise to three phenotypes: skeletal, pul-monary, and skin. The PTHrP ligand and receptor, and their amplified signaling through the second messengers cyclic AMP and inositol triphosphate are common to all of these disparate processes, whereas the downstream target genes vary, giving rise to a variety of lung cell phenotypes. Similarly, insulinlike growth factor (IGF), platelet-derived growth factor (PDGF), fibroblast growth factor (FGF), transforming growth factor beta (TGFβ), and other growth factors all signal through receptors on various cell types, differing with regard to what structural and functional phenotypic genes they signal to in those target cell types.

This pattern of physiologic effects mediated by signaling molecules from one cell type stimulating neighboring cell types through a specific receptor-mediated recognition mechanism is the basis for the evolution of multicellularlity. This variation in cell surface markers identifies the signaling partners in a recapitulation of evolution that plays itself out iteratively through ontogeny and phylogeny, and reprises itself during injury repair.

EVOLUTION AS THE SOLUTION

We are just beginning to recognize the convergent expression of genes that have given rise to the complexity of physiologic processes. Lung evolution has been driven by amplification of the PTHrP signaling pathway, which, in turn, has pro-vided insights into the development, function, and dysfunction of the lung. Thus, the phylogeny, ontogeny, physiology, and pathophysiology of the lung may be expressed as a family of parallel lines (see Chapter 3), providing enough variables to solve for all these simultaneous equations, but perhaps having to skip around from one line to another for lack of a complete dataset for any given line in the family.

The solution for this set of simultaneous equations for the lung is the process of evolution. The other PTHrP-dependent phenotypes, bone and skin, could be evaluated similarly, providing for a systematic approach to the integration of biologic knowledge. By using interactive algorithms such as self-organizing maps or neural nets to integrate the totality of biologic data, the periodic table for biology will emerge (Fig. 8.1), based on first principles, rather than on description and teleology. Mendeleev's Periodic Table gave chemists an algorithm to understand the behavior of the elements. If the biologic version of the periodic table is as robust and predictive, it will provide novel ways in which to view life.

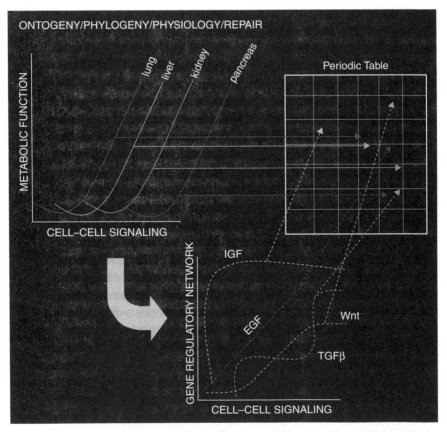

Figure 8.1. The biologic periodic table. In the upper left-hand corner, cell–cell signaling for the ontogeny, phylogeny, and repair of various organs is regressed against metabolic function. Below, the descriptive data are replaced by mediators of the signaling pathways. To the right, the periodic table of biology is depicted as a four-dimensional hologram. (See insert for color representation.)

RAMPING UP A MATHEMATICAL MODEL OF EVOLUTION

In Chapter 7 we suggested that by starting with the Horowitz retrograde evolution of biochemical pathways, and superimposing the cell–cell interactions that facilitated the genetic changes, we could deconstruct and reconstruct the evolutionary mechanisms involved in the generation of physiology (Fig. 8.2). Since we would ultimately like to create a mathematical expression for this process, we have taken the first steps toward conceptually merging the schematics for cellular, genomic, and physiologic evolution (Fig. 8.2); then, by assigning algebraic notation to these processes (Fig. 8.3), we can begin generating a mathematical model. The progression from associative (x,y), to additive $(x + y)$, to multiplicative (xy), to exponential (xy^e), to combinatorial/permutative $\{xy^e\}!$ relationships is based largely on the evolutionary cell–cell interaction model for the lung alveolus, in which there is initially merely an association (x,y) between connective tissue and epithelium in the swim bladder, followed by additive housekeeping relationships $(x + y)$ in the amphibian lung, to simple *cis* regulatory multiplicative relationships (xy) in the reptilian lung, to complex exponential *cis* regulatory relationships (xy^e) in the mammalian and avian lung, to combinatorial/permutative *cis* regulatory relationships $\{xy^e\}!$ between the lung and other organ systems (heart, adrenal, adipose, kidney, liver).

As mentioned above, the power of the Mendeleev Periodic Table was in the use of atomic number as the organizing principle for the hierarchy, in conjunction with valence. By analogy to the effect of the cosmic Big Bang on the generation and distribution of the physical elements, compilation and construction of a biologic periodic table might be based on a cholesterol hierarchy, given its centrality to the Big Bang of vertebrate evolution by biochemically linking its synthesis to atmospheric oxygen. The hierarchic position and valence for any given component of this table would be a function of its structural and/or functional molecular homology to cholesterol, as an example of a prototype, determined by a database search using such bioinformatics search engines such as PubMed, KEGG (*Kyoto Encyclopedia of Genes and Genomes*), Chilibot, GeneSpring, or the Basic Local Assignment Search Tool (BLAST). The more proximate the biologic element is to a cholesterol homolog, the higher, and to the left on the table it would rank, by analogy with the Mendeleev table of elements. Such a table would, for example, predict key cellular–molecular events in the evolution of the lung, and that of other tissues and organs, as depicted in Chapter 7 (Fig. 7.5). As a result, it would resolve the many paradoxes of biology which have been highlighted during the course of this book, generating a unifying theory of biology, and a rational basis for predictive medicine. Even more importantly, such a periodic table for biology could be merged with the table of elements, forming a common database for physics, chemistry, and biology, as proposed by E. O. Wilson (see introduction to Chapter 4).

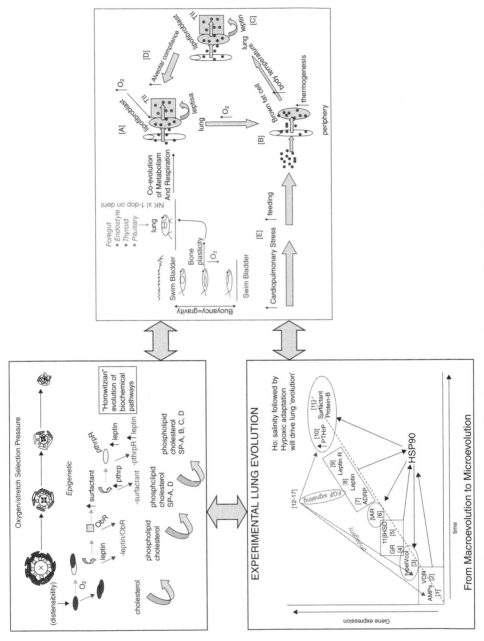

Figure 8.2. Merging schematics for cellular, genomic, and physiologic evolution. The schematics from earlier chapters are conceptually merged together. (See insert for color representation.)

122

How Genes Recombine to Generate Novel Phenotypes

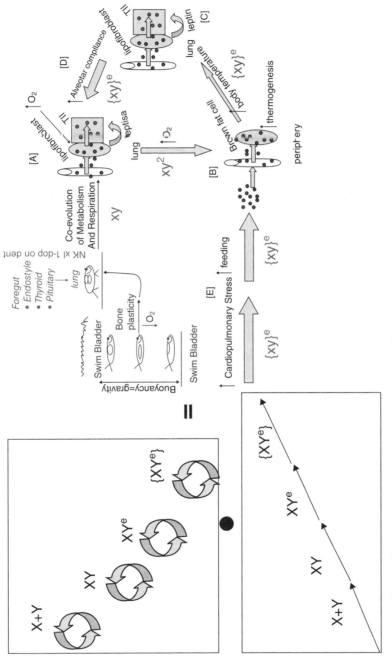

Figure 8.3. Assignment of algebraic expressions for the evolutionary integration of cellular, genomic, and physiologic evolution. In anticipation of mathematical expression of the evolution model, we have assigned algebraic characters for semiquantitation. (See insert for color representation.)

123

THE ANTHROPIC PRINCIPLE RESULTS FROM THE
EVOLUTION OF CELL–CELL INTERACTIONS

The Anthropic Principle was formulated in response to observations that the laws of nature and its physical constants acquire values that are remarkably consistent with conditions for life as we know it on Earth. This apparent coincidence is actually necessary, since living observers wouldn't be able to exist and perceive the universe if these laws and constants were not consistent with life on Earth. But this begs the question as to why we are "at home in the Universe," as Stuart Kaufmann suggests. The seeming paradox of the Anthropic Principle results from reasoning after the fact. Instead, we would like to suggest that the Anthropic Principle is the result of the cell–cell interactions that have mediated the biologic adaptation to the physical environment. In other words, *it is because we have evolved in response to the environment that we ascribe to the physical principles that determine it.*

The term *Anthropic Principle* first appeared in a speech by the theoretical astrophysicist Brandon Carter at a symposium in 1973 in Krakow, Poland celebrating Copernicus' 500th birthday. Carter invoked the Anthropic Principle in response to the Copernican principle that humankind is not at the center of the universe. Carter, in turn, argued that even though humans are not at the center of the universe, we do occupy a place of privilege. In particular, Carter took exception to the use of the Copernican principle to substantiate the "perfect cosmological principle," which claims that all large regions and times in the universe must be statistically identical.

The Anthropic Principle is based on the underlying belief that the universe was created for human benefit. Unfortunately for its adherents, all of the reality-based evidence at our disposal contradicts this belief. In a nonanthropocentric universe there is no need for multiple universes or supernatural entities to explain life as we know it.

Similarly, Stephen Jay Gould, Michael Shermer, and others have claimed that the Anthropic Principle seems to reverse known causes and effects. Gould facetiously compared the claim that the universe is fine-tuned for the benefit of our kind of life to saying that sausages were made long and narrow so they could fit into hotdog buns, or that ships had been invented to house barnacles. These critics cite the vast physical, fossil, genetic, and other biological evidence consistent with life having been fine-tuned through the process of natural selection to adapt to the physical and geophysical environment in which life exists. Life adapted to physics, and not vice versa. We take exception to the invocation of natural selection because it is descriptive, whereas the effects of the physical world on cell–cell interactions are adaptive, mechanistic, and predictive.

In Chapter 9 we will highlight the advantages to the cellular–molecular approach to physiologic evolution.

9

VALUE ADDED BY THINKING IN TERMS OF THE CELL–CELL COMMUNICATION MODEL FOR EVOLUTION

In this book, we have shown that by focusing on the cell as the smallest functional unit of biology, and by following the progression from unicellular to multicellular organisms, the process of evolution can be seen clearly. As a result, this approach provides a predictive and testable model for the evolution of biologic structure and function.

Evolutionary Biology, Cell–Cell Communication, and Complex Disease, First Edition.
John S. Torday and Virender K. Rehan.
© 2012 Wiley-Blackwell. Published 2012 by John Wiley & Sons, Inc.

THIS IS *NOT* A JUST-SO STORY

Unlike Kipling's just-so stories about how and why the leopard got its spots, the rhinoceros got its tough skin, or the camel got its hump, the cell–cell signaling model of physiologic evolution does not reason after the fact. It is based on evolved mechanisms of cell–molecule embryogenesis, culminating in homeostasis. When such evolved structure–function relationships fail, they recapitulate developmental–evolutionary mechanisms. Unlike the classic pathophysiologic approach to disease, which reasons backward from disease to health, the evolutionary–developmental approach, like Figure 4.1, reasons from the cellular origins of physiology, resulting in prediction of the cause of chronic disease, as we have shown for the lung, and Leon Fine has shown for the kidney (Bacallao and Fine 1989). As proof of principle, we will cite three examples of the fundamental difference between a pathophysiologic versus an evolutionary approach to disease.

Lung Prematurity and Bronchopulmonary Dysplasia

On the basis of our own work, we have been able to effectively predict and prevent a chronic lung disease of the newborn, namely, bronchopulmonary dysplasia (BPD), experimentally, according to principles of pulmonary cellular–molecular evolution. Unlike the conventional way of viewing chronic lung diseases such as BPD as the result of inflammation, BPD is the result of prematurity, independent of inflammation—simply overdistending the preterm lung using a mechanical ventilator, or exposing it to room air can cause BPD.

By focusing on the epithelial–mesenchymal interactions that normally form the lung during development, we have found that those infants that develop BPD are PTHrP-deficient. Additionally, by specifically stimulating peroxisome proliferator–activated receptor gamma (PPARγ) in the lung fibroblast, which is the PTHrP target that drives alveolar development and homeostasis, we can prevent BPD.

Osteoporosis and the Manned Space Program

Similarly, the etiology of osteoporosis may be better understood using the evolutionary–functional genomic approach. On the basis of conventional pathophysiology, because the bone–wasting disease osteoporosis occurs commonly in postmenopausal women, its cause has been attributed to decreased estrogen production. Ironically, this same disease occurs in male astronauts, who clearly do not suffer from decreased estrogen production, yet the disease is being attributed to this same endocrine mechanism. On the contrary, we have implicated microgravitational effects of PTHrP in this process, based on an evolutionary adaptational mechanism, whereby the tension of muscle on bone causes locally increased PTHrP production, maintaining calcium homeostasis. This evolutionary interrelationship between mechanotransduction and bone calcification is consistent with studies of calcification in atherosclerosis. Here, too, signaling for calcium homeostasis may come into play when the normal homeostatic mechanisms fail, perhaps as an adaptive means of maintaining vascular bloodflow.

Myelinization and Diseases of the Central Nervous System

As an example of how an evolutionary approach can facilitate understanding of complex neurologic diseases, Cochran et al. (2006) hypothesized that the high intelligence test scores among Ashkenazi Jews are a consequence of their social isolation from the general population in shtetls over much of the last millennium that selected strongly for intelligence. The investigators went on to suggest that there was an increase in the frequency of specific genes that caused this increased intelligence, which also led to an increased incidence of hereditary neurologic disorders characterized by abnormal neuronal myelinization. The mechanisms that favored increased neuralization for intelligence inadvertently gave rise to such neurodegenerative diseases as Tay–Sachs, Gaucher's, Niemann–Pick, and mucolipidosis type IV, all of which are lysosomal storage diseases. The increase in storage of glucosylceramide in Gaucher's, for example, promotes axonal growth and branching. Similarly, in Tay–Sachs and Niemann–Pick diseases, increased GM2 ganglioside causes increased dendritogenesis.

Interestingly, such a mechanism would mechanistically link the balancing selection pressure for vitamin D to its effect on skin color, as mentioned in Chapter 5, since all of the neurodegenerative diseases listed above are associated with skin abnormalities due to the lipid storage defect. Lipids in the stratum corneum of the skin form a physical barrier in combination with β-defensins. Therefore, selection pressure for this deep homology between host defense, lipid metabolism, and brain myelinization may, on one hand, have facilitated migration early in human history but also precipitated both neurodegenerative and skin diseases, albeit in a small subset of people, on the other hand. In addition, as indicated in Chapter 7, leptin has facilitated these processes. Perhaps that is why substrate-lowering agents used to treat hyperglycemia or hypercholesterolemia have been found to be effective in treating neurodegenerative diseases. More importantly, rather than treating such diseases for their overt metabolic abnormalities, perhaps using an evolutionary approach to address the deep homologies would be more effective. Pharmacologic amounts of vitamin D may activate innate host defense mechanisms to counter the inflammatory effects of excess lipid production. For example, it has been shown that vitamin D treatment reduces the incidence of schizophrenia, and vitamin D deficiency is associated with Gaucher's disease.

BEYOND GENOMICS

Three thousand years of descriptive biology and medicine have brought us to the threshold of molecular medicine. Now, aided by our knowledge of the human genome, we must address the evolutionary origins of human physiology On the basis of phylogenetic and developmental mechanisms. The cellular–molecular approach that we have proposed may fail to directly identify such first principles because we are missing intermediates from the molecular fossil record that failed to optimize survival. But some vestiges of those failures were likely to have been incorporated into other existing functional phenotypes, or into other molecularly related functional homologies, such as those of the lung, the kidney, the brain, photoreceptors, and circadian rhythms, the lens and liver enzymes. What this approach does provide is a robust means of formulating refutable hypotheses to determine the ultimate origins and first principles of physiology by providing candidate genes for phenotypes hypothesized to have mediated evolutionary changes in structure and/or function. It also forms the basis for predictive medicine rather than merely showing associations between genes and pathology, which is an unequivocal just-so story. In the new age of genomics, our reach must exceed our grasp.

FROM FAT CELLS TO INTEGRATED PHYSIOLOGY

We have previously addressed the developmental, homeostatic, regenerative, and evolutionary roles of the lipofibroblast by identifying key GRNs relevant to the evolution of the lung—namely, the PTHrP/leptin GRN that determines the development, homeostasis, and regeneration of the lung alveolus. By focusing on this key paracrine mechanism, we have been able to identify genes expressed in lipofibroblasts upstream from the surfactant mechanism that are highly conserved in vertebrate physiology, namely, PPARγ, ADRP, and leptin. The stretch regulation of lung surfactant by PTHrP likely refers back to the swim bladder origins of the lung, since the swim bladder integrates the physical adaptation to gravity (i.e., buoyancy) with feeding efficiency. The relevance of PTHrP to lung evolution was first suggested by the lung phenotype in the PTHrP knockout mouse, since the absence of PTHrP resulted in failed alveolar formation, suggesting its principal role in lung evolution. The evolution of the lung is characterized by the progressive phylogenetic decrease in alveolar diameter, increasing the surface area for gas exchange from fish to humans. That progression would not have been physically possible without the accompanying increase in the efficiency of surfactant synthesis to reduce alveolar surface tension, since the law of Laplace dictates that the smaller the surface area of a sphere, the higher its surface tension.

In retrospect, experimental evidence for the evolutionary interrelationship between surfactant and stretching was first provided by Clements et al. (1970), when they demonstrated a linear interrelationship between alveolar surface area and the amount of surfactant per unit area across a wide range of vertebrate species, ranging from frogs to cows. In the interim, there have been numerous experiments demonstrating the on-demand production of surfactant in response to increased tidal volume, namely, alveolar distension, culminating in a series of studies from our laboratory showing how the stretching of the alveolus determines its physiological adaptation and plasticity via the PTHrP–leptin signaling pathway.

The evolutionary significance of this mechanism is further underscored by the pleiotropic effects of PTHrP and leptin on other physiologically integrated principles of the alveolus: the potent vasodilatory effect of PTHrP-facilitated alveolar ventilation perfusion matching, the physiologic integration of alveolar expansion, and contraction with alveolar capillary perfusion, further enhancing the efficiency of gas exchange by coordinating lung tidal volume with both surface-tension-reducing activity and alveolar capillary bloodflow (V/Q matching). Also, the evolved regulation of type IV collagen is indicative of the physiologic monitoring of the blood–gas barrier, as first suggested West and Mathieu-Costello (1999).

MOLECULAR HOMOLOGIES DISTINGUISH BETWEEN THE EVOLUTIONARY FOREST AND TREES

The central premise of this book is that there are molecular homologies that connect seemingly disparate physiologic phenotypes. Those homologies can be visualized through the prism of developmental and phylogenetic cellular–molecular mechanisms. For example, we have been able to reduce metabolic adaptation of terrestrial vertebrates to the induction of adipocytes by oxygen in both the lung and in the periphery. The presence of these stored neutral lipid droplets acts to protect the alveolus against oxidant injury. Moreover, the fat cell secretory product leptin has pleiotropic effects on the development of the lung, vasculature, bone, and the central nervous system, providing a mechanistic link between adipocytes and the evolution of complex physiologic structures and functions. Empiric evidence for these mechanistic interrelationships comes from the mouse knockout for peroxisome proliferator–activated receptor gamma (PPARγ), the nuclear transcription factor that determines the adipocyte phenotype, generating both fat cells and lung lipofibroblasts. As further evidence for the role of leptin in normal lung development, the leptin-deficient ob/ob mouse lung is hypoplastic, and this deficiency can be corrected by treating the mice with leptin.

Our ability to test the hypothetical linkage between adipocyte production of leptin and the evolution of complex physiologic traits has more recently been further expanded by the discovery of micro-RNAs, which are endogenously expressed 20–24-nucleotide RNAs thought to repress protein translation through binding to target messenger RNAs. Only a few of the more than 250 predicted human micro-RNAs have been assigned any biological function. Micro-RNA (miR)-143 levels increase during adipocyte differentiation, and inhibition of miR-143 effectively inhibits adipocyte differentiation. In addition, protein levels of the proposed miR-143 target ERK5 are affected by miR-143 antisense oligonucleotide-treated adipocytes, demonstrating the causal relationship between miR-143, ERK5, and adipocyte differentiation. Both miR-103 and miR-107 have also been shown to play roles in adipogenesis.

THE OXYGEN–CHOLESTEROL–SURFACTANT–MEMBRANE CONNECTION

The interrelationship between atmospheric oxygen, cholesterol, and plasma membrane evolution, and how it facilitated the evolution of eukaryotes from prokaryotes by allowing for a thinner plasma membrane for gas exchange and cytosis, was discussed at length in Chapter 1. We have linked these evolved traits together later in vertebrate evolution (see Fig. 9.1) through the effects of lung surfactant on the fish swim bladder, leading to the evolution of the lung in amphibians, reptiles, and mammals. For example, even though there are similarities between apolipoproteins facilitating the metabolism of circulating lipids, and the effect of ADRP on triglyceride uptake by alveolar type II cells, deletion of the cholesterol synthesis gene Scap from the alveolar type II cell leads to compensatory overexpression of the lipofibroblast phenotype, providing more lipid substrate for surfactant synthesis, recapitulating the evolved response to surfactant deficiency. In contrast to this, a deficiency in apoprotein E (APOE) cholesterol transport causes lung fibrosis, as reported by Massaro and Massaro (2008), representing failed compensation, or a

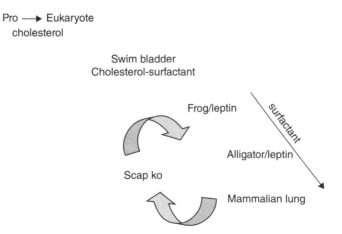

Figure 9.1. Depiction of the phenotypic interrelationship between cholesterol, lung surfactant, and the evolution of the lung. Beginning in the upper left corner, cholesterol facilitated the evolution of prokaryotes to eukaryotes, followed by lung evolution through cell–cell interactions that determined surfactant regulation.

maladaptation. This fundamental difference in the biologic response to these two nominal cholesterol regulatory genes reflects the fundamental difference between the loss of a housekeeping gene (APOE) versus the deep homologies being elicited by the loss of a *cis* regulatory gene (Scap). To illustrate further, the Scap deletion also caused increases in genes for endoplasmic reticulum stress, which gave rise evolutionarily to peroxisomes, interrelating selection pressure for increases in both cholesterol for endoplasmic reticulum membrane expansion (see Chapter 1), and PPARγ, the determinant of the lipofibroblast phenotype.

As an example of how the cellular–molecular evolutionary approach predicts physiologic interrelationships, Krogh (1942) recognized the central importance of lung architecture for oxygen demand, supported by the presence across the entire range of mammalian body sizes of a direct linear relationship between alveolar surface area and oxygen uptake. This matching of body size to lung physiology has been achieved by progressive subdivision of the alveolar surface, causing smaller and smaller alveoli, increasing the blood volume/surface area, ratio of the lung for increased gas exchange. This highly conserved interrelationship between alveolar size, alveolar number, surface area, and oxygen exchange indicates the existence of an underlying mechanism that determines the overall process of alveolarization. As a result, the organism's oxygen demand is achieved at a breathing rate and tidal volume for which the work of breathing is most efficient. The lipofibroblast, which likely evolved to protect the lung in response to the rising oxygen tension in the atmosphere (see Chapter 2), also produces leptin, a pleiotropic hormone that stimulates epithelial cell growth, host defense, surfactant production, and the rate of breathing. ob/ob mice genetically deficient in leptin have small lungs; treating them with leptin increases both their alveolar surface area and their rate of breathing, demonstrating the evolutionarily adaptive, multifunctional roles of the metabolic hormone leptin.

There is mounting evidence for the overall importance of cholesterol in vertebrate physiology and disease. Cholesterol affects fetal development by facilitating sonic hedgehog signaling, which is key to morphogenesis. In food restriction models of altered fetal and postnatal development, cholesterol metabolism is affected in a stage- and sex-dependent manner, which is consistent with the well-known associations between cholesterol and coronary heart disease, obesity, and hypertension. All of these observed relationships between cholesterol, physiology, and pathophysiology are indicative of the seminal role played by cholesterol in vertebrate evolution.

CHOLESTEROL METABOLISM AS THE DATA OPERATING SYSTEM FOR VERTEBRATE BIOLOGY?

We have stipulated that the process of evolution is cell communication at multiple levels—unicellular organisms communicating with their environment; cell–cell communication as the basis for the evolution of multicellular organisms; and ultimately, cell communication of genetic information from one generation to the next, or reproduction. We can estimate the value added in thinking about cholesterol metabolism as the central mechanism driving vertebrate evolution by looking at its role in each of these processes. For example, cholesterol was critical in the evolution from prokaryotes to eukaryotes, facilitating cytosis by the plasma membrane (see Chapter 1). Cholesterol is also important in signal transduction since it generates lipid rafts in the plasma membrane. These lipid rafts facilitate receptor-mediated communication with the environment, and then between cells as cell–cell communication, which is the basis for the evolution of multicellular organisms. Furthermore, cholesterol integrates tissue homeostasis in lung, liver, kidney and other, organs. Ultimately, cholesterol is critically important for endocrine regulation of homeostasis and reproduction, since it is the substrate for steroid hormones such as estrogen, progesterone, and androgen that orchestrate sexual reproduction. This strategy has now been mechanistically linked with the overall lifecycle, since food deprivation during the latter half of pregnancy programs the metabolism of the offspring. Moreover, it is now known that cholesterol affects hedgehog signaling, which is essential for normal embryogenesis, providing a cellular–molecular basis for the role of cholesterol in determining all of the milestones in biology—development, homeostasis, regeneration, and reproduction.

At first glance, the notion that cholesterol metabolism is the central data operating system for vertebrate biology may seem to be an oversimplification, given the obvious complexity of this process. Another way of phrasing this is to ask: Why are lipids the key class of compounds that have facilitated evolution? The answer may lie in the origins of life as the reduction of entropy caused by the formation of semipermeable lipid spheres, or micelles, as discussed in Chapter 1. Other classes of biologic substances such as nucleotides, proteins, or carbohydrates do not have the capacity to facilitate membrane formation, and were more likely to have been generated as a result of evolutionary selection pressure. Furthermore, lipids represent a wide array of biologic properties, ranging from the structure of cell membranes, to organellar membranes, to signaling molecules and antimicrobial activity. The cell membrane is the fundamental structure of biology, forming the boundary between the external and internal environments. Seen from this perspective, there is a sort of resonance throughout the form and function of vertebrate biology mediated by the evolution of cholesterol.

TRANSLATION OF GENOMICS INTO THE PERIODIC TABLE FOR BIOLOGY

Our evolutionary–developmental paradigm for lung biology can be expanded to a systems–biology-like approach through a number of avenues (see Fig. 4.2).

Embryologically, as we have already indicated, the PTHrP signaling pathway is up-stream from the endocrine mechanisms that determine development of the lung, incorporating the endocrine system into the model. The coevolution of the lung and endocrine systems is also functionally linked through the physiologic adaptation of cholesterol for surfactant activity in the case of the lung, and as substrate for steroid hormone synthesis in the case of the endocrine system. Selection pressure for that interrelationship may well have come from the episodic increases and decreases in atmospheric oxygen over the last ~500 million years, generating huge selection pressures for adaptation to both hyperoxic and hypoxic conditions, especially when bearing in mind that 99.9% of all animal life became extinct over this period. Furthermore, because the lung is embryologically derived from the gut, it develops through similar gene regulatory pathways, thus developmentally integrating the gut and its derivatives—the liver, pancreas, thyroid, and pituitary—into the model. These functional phylogenetic and developmental interrelationships can be further exploited by reducing them to their molecular phenotypes. For example, we have demonstrated the centrality of both the Wnt/β-catenin and G-protein-coupled protein kinase A pathways in the epithelial–mesenchymal interactions that determine pulmonary structure and function. These same epithelial–mesenchymal interactions determine the structure and function of a broad variety of tissues and organs (kidney, liver, pancreas, gut, thyroid, adrenal, thymus, eye, skin). Also, the central role of peroxisome proliferator–activated receptor gamma (PPARγ) as the determinant of the lung lipofibroblast gene regulatory network (GRN) mechanistically engenders such seemingly disparate tissues and organs as the brain, bone marrow, liver, and adipose tissue, all of which are also phenotypically Peroxisome Proliferator PPARγ-dependent.

The networks of genes that derive from the periodic table approach proposed in Chapter 8 would link specific Phenotypes Together in ways that seem counter-intuitive, offering dynamic new ways of thinking about how the genomic elements (components) of physiologic systems are recombined and permuted through the process of evolution to generate novelty based on cellular principles of phylogeny and development, rather than on post-dictive descriptions of structure and function. This is analogous to the Periodic Table of Elements being based on atomic number as an independent organizing principle for the physical elements. Also like the Periodic Table of Elements, which predicted the existence of additional elements, the biologic algorithm will predict novel molecular phenotypes and gene regulatory networks. Ultimately, this biologic algorithm will reveal the underlying rules for the first principles of physiology. Our laboratory has devised several models with which to test this evolutionary cellular–molecular concept—the developing rat and mouse, the embryonic chick and alligator, and the frog tadpole. These models offer a concerted developmental and phylogenetic approach for determining specific functional gene regulatory networks across phyla.

DEEP HOMOLOGIES

Because of the profound similarities in genetic regulation between organs as disparate as fly wings and tetrapod limbs, Shubin et al. (2009) have suggested the term "deep homology" to describe the sharing of genetic regulatory apparatuses that are used to build morphologically and phylogenetically varied animal traits. Such deep homologies are the key to the generation of novelties because such ancient regulatory circuits for cell physiology provide the means for novel structures to develop. Using such classic examples of evolution as the eye, fin, limb, and heart, for each of these structures, the same gene circuits are utilized, all the way back to early metazoans and unicellular organisms (see Chapter 1). Sean Carroll (2008) has gone on to emphasize the importance of the evolution of *cis* regulatory elements for DNA as the basis for the mechanism of evolution. The concept that evolution must be mediated by changes in the regulation of genetic expression was triggered by King and Wilson (1975), who first pointed out that although humans and chimpanzees are genetically almost identical, they show profound phenotypic differences, raising questions about the fundamental nature of genetic mechanisms of evolution yet again. The subsequent discovery of homeobox genes that determine segmentation as being common to flies and vertebrates alike led yet again to developmental geneticists having to confront evolutionary issues, and evolutionists having to deal with the genetics of animal form. These interfaces gave rise to evolutionary–developmental biology, or evo-devo. Many examples of genetic circuits forming the basis for the evolutionary process have been documented, ranging from Pax6 in the eye to antennapedia and distalless in limbs, yet the linkage to *cis* regulatory elements, or how natural selection affected them to evolve structure and function, is lacking. Hoekstra and Coyne (2007) point out that even in the case of well-accepted examples of molecular evolution, none have demonstrated an evolved regulatory site.

In contrast to this, the model of lung evolution that we have proposed, based on the historic and developmental progression of cell–cell interactions that generate lung surfactant, provides a means of systematically identifying the intermediate *cis* regulatory changes that evolved. We have pointed out that lung evolution is characterized by the progressive regulation of housekeeping genes; that process is dependent on the emergence of novel *cis* regulatory mechanisms. To illustrate, we will recapitulate the ontogenetic changes in the lung (see Fig. 9.2), pointing out the acquisition of *cis* regulatory elements as we go along: PTHrP of epithelial cell origin binds to its receptor on neighboring lung fibroblasts, stimulating cyclic AMP-dependent protein kinase A. This pathway specifically induces PPARγ and its downstream targets ADRP and leptin through various *cis* regulatory mechanisms to form the lipofibroblast. Leptin produced by the lipofibroblast binds to its receptor on the epithelial cell, stimulating phosphatidylInositol 3 kinase, which induces *cis* regulation of surfactant protein and phospholipid, retinoid X receptor (RXR), as well as the production of PGE_2, which is necessary for the release of stored lipid droplets from the lipofibroblast for surfactant phospholipid synthesis.

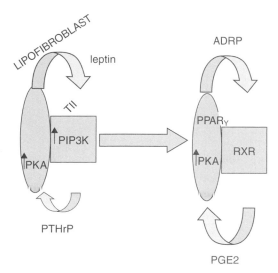

Figure 9.2. Lung evolution from housekeeping genes to *cis* regulation. PTHrP signals through cAMP, inducing PPARγ, ADRP, and leptin. Leptin signals through phosphatidylinositol 3 kinase (PIP3K) to stimulate surfactant protein and phospholipid, retinoid X receptor (RXR), and PGE$_2$.

These lipid droplets contain retinoic acid, which is necessary for the maintenance of both the lipofibroblast and epithelial cell phenotypes. Retinoic acid binds to its nuclear receptors, further acting on a variety of *cis* regulatory elements within the alveolus to maintain homeostasis. These *cis* regulatory mechanisms are common to the development and homeostasis of the lung, gut, liver, skin, bone, and kidney, providing a molecular cipher for reverse-engineering the evolution of these tissues and organs. In addition, since these organs have also evolved through various internal and external adaptive selection pressures (pointed out in Chapters 5 and 6), the underlying *cis* regulatory mechanisms can also be based on the basis of those criteria, using ontogeny and phylogeny as hierarchical guides and reference points.

SELECTION PRESSURE FOR CELL–CELL COMMUNICATION: THE KEY TO UNDERSTANDING EVOLUTION

In the course of the first eight chapters of this book, we have built on the concept of cell–cell communication as the integrating principle for selection pressure and the evolution of complex physiologic traits. This mechanism integrates physiology from molecular oxygen to homeostasis (see Fig. 9.3), providing a logic for the form and function of specific organs such as the lung and thyroid, to whole-animal physiology (see Chapter 7). Others have described various elements of such a

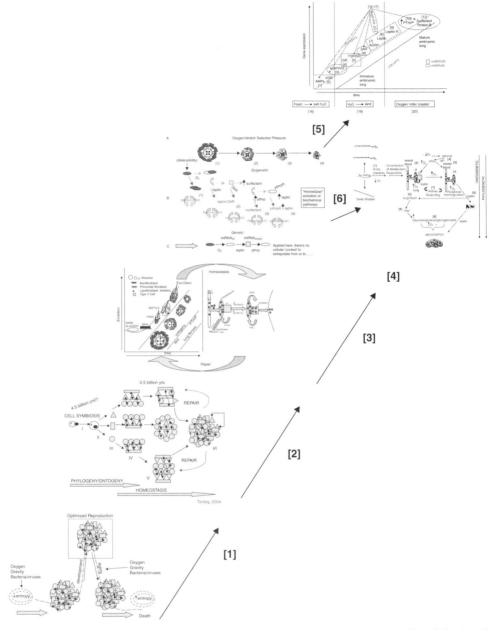

Figure 9.3. Integration of physiology. Continuum from [1] the advent of the cell, to [2] cell–cell signaling for multicellularity, [3] lung cell evolution based on surfactant production, [4] evolution of *cis* regulation, [5] intrinsic–extrinsic selection pressure, and [6] the lung as a cellular–molecular basis for physiologic evolution. (See insert for color representation.)

mechanism of evolution, focusing on specific components such as the primordial origin of cells, the evolution of multicellular organisms, commonalities between ontogeny and phylogeny, and selection pressure and its influence on *cis* regulatory mechanisms. We have chosen to refer to these founding principles as the biologic *data operating system*.

The present model is unique among the myriad explanations of vertebrate evolution because it provides a way of thinking about the entire process of evolution from microevolution to macroevolution as a mechanistic continuum—and of equal importance, it provides an experimentally testable set of hypotheses. The novelty of this approach derives from the focus on molecular phenotypes that originated in unicellular organisms communicating with the environment, forming a heritable, stable cellular–molecular platform on which to build multicellular organisms, based on the founding unicellular principles (as represented by the schematic in Fig. 9.3). The evolution of multicellular organisms from unicellular organisms occurred as a result of cell–cell communications progressively modifying *cis* regulatory mechanisms through molecular second messengers such as cyclic AMP, phosphatidyl inositol, and retinoic acid that mediate the effects of such cell–cell communication mechanisms by modifying *cis* regulation. Therefore, the origins of evolved traits can be determined by analyzing the changes in the *cis* regulatory mechanisms within the context of the phenotypic structural and functional ontogeny and phylogeny of any particular trait, as illustrated in Figure 9.3.

In the case of lung evolution, for example, the epithelial lining cells of the swim bladder express surfactant genes as a housekeeping function, but are not regulated through ligand–receptor-mediated cell–cell interactions. In mammals, on the other hand, the surfactant proteins and lipids are highly regulated by paracrine mechanisms that integrate the stretching of the alveolar wall with its capillary perfusion, referred to as V/Q matching. The paracrine factors that mediate these physiologic effects, PTHrP and leptin, are mutually produced by alveolar epithelial cells and their underlying interstitial fibroblasts, respectively, and the specific cell surface receptors for these signaling proteins are likewise present on neighboring fibroblasts and epithelial cells. When these paracrine factors bind to their receptors, they coordinately generate molecular messengers that regulate surfactant genes in the epithelial cells, and surfactant-related functions in the fibroblasts by modifying the surfactant-related *cis* regulatory mechanisms within the nuclei of these cells. As experimental evidence, Brigid Hogan's laboratory (Okubo and Hogan 2004) has shown that molecularly suppressing lung development results in the expression of intestinal epithelial cells, reflecting the evolutionary origin of the lung. These cellular–molecular communication properties have evolved under selection pressure for optimal gas exchange over biologic time from fish to mammals, and by methodically analyzing them within and between vertebrate species, using the constitutive and regulated states of surfactant production as the functional molecular phenotypes, the underlying evolutionary mechanisms can be sorted out.

Moreover, in Chapter 5 we showed how other organs coevolved with the lung through selection pressure for oxygenation, beginning with the adrenal glands, followed by the fat cell depots in the lung and periphery: the liver, kidney, and brain. This approach allows for the expansion of this systematic contextual analysis of *cis* regulatory evolution based on the molecular functional integration of these related physiologic properties. That analysis can be further refined by calibrating the *cis* regulatory mechanisms of interest with the intrinsic and extrinsic genetic mechanisms of lung evolution, and with that of other organs. Indeed, the lung could be viewed as the contingency on which the emergent physiologic properties were selected. Such fine tuning of vertical integration is important because it provides the rules and exceptions, both positive and negative, for determining the unequivocal mechanisms of physiologic evolution.

In Chapter 10 we will show how the cell–cell signaling approach to evolution can be applied to medicine. By reducing physiology to its first principles, medicine will become predictive.

REFERENCES

Bacallao R, Fine LG (1989), Molecular events in the organization of renal tubular epithelium: from nephrogenesis to regeneration. *Am. J. Physiol.* 257(6 Pt. 2):F913–F924.

Carroll SB (2008), Evo-devo and an expanding evolutionary synthesis: a genetic theory of morphological evolution. *Cell* 134(1):25–36.

Clements JA, Nellenbogen J, Trahan HJ (1970), Pulmonary surfactant and evolution of the lungs. *Science.* 169(945):603–604.

Cochran G, Hardy J, Harpending H (2006), Natural history of Ashkenazi intelligence. *J. Biosoc. Sci.* 38(5):659–693.

Hoekstra HE, Coyne JA (2007), The locus of evolution: evo devo and the genetics of adaptation. *Evolution.* 61(5):995–1016.

King MC, Wilson AC (1975), Evolution at two levels in humans and chimpanzees. *Science.* 188(4184):107–116.

Krogh A (1942), *Comparative Physiology of Respiratory Mechanisms.* University of Pennsylvania, Philadelphia, PA, p. 3.

Massaro D, Massaro GD (2008), Apoetm1Unc mice have impaired alveologenesis, low lung function, and rapid loss of lung function. *Am. J. Physiol. Lung Cell. Mol. Physiol.* 294(5): L991–L997.

Okubo T, Hogan BL (2004), Hyperactive Wnt signaling changes the developmental potential of embryonic lung endoderm. *J. Biol.* 3(3):11.

Shubin N, Tabin C, Carroll S (2009), Deep homology and the origins of evolutionary novelty. *Nature* 457(7231):818–823.

West JB, Mathieu-Costello O (1999), Structure, strength, failure, and remodeling of the pulmonary blood-gas barrier. *Annu. Rev. Physiol.* 61:543–572.

10

CELL–CELL COMMUNICATION AS THE BASIS FOR PRACTICING CLINICAL MEDICINE

Evolutionary Biology, Cell–Cell Communication, and Complex Disease, First Edition.
John S. Torday and Virender K. Rehan.
© 2012 Wiley-Blackwell. Published 2012 by John Wiley & Sons, Inc.

CELL–CELL COMMUNICATION MAINTENANCE AND BREAKDOWN REPRESENT HEALTH AND DISEASE, RESPECTIVELY

In previous chapters, by interrelating genes, molecules, intercellular-communications, and structural and functional phenotypes in their historic contexts, we have provided a framework for deconvoluting the first principles of biology and human evolution in the process. We have shown how, in the face of external and internal stresses—such as exposure to oxygen, infection, dietary, and other environmental challenges—the contemporary human lung and other organs have evolved over eons. In this chapter, we will show how these concepts and principles predict the genetic, molecular, cellular, structural, and functional phenotypes of the contemporary lung, and how they can be exploited in clinical medicine. The overarching theme of this approach is that the health of an individual is not merely the absence of disease but that it is the active maintenance of evolutionarily acquired and conserved cell–cell molecular communications, the failure of which denotes disease, and the restoration of which leads to healing and health.

The examples cited and the approach outlined in this chapter provide a novel way of thinking about clinical issues; specifically, the body's response to disease can be predicted on the basis of historic cellular responses to positive selection pressures during each organ's development. For example, given the signaling pathway commonalities from the evolution of the fish swim bladder to the mammalian lung, it is predicted that the genes in this particular paracrine pathway will be highly polymorphic, which they are, providing plasticity for physiologic evolution, on one hand, and a basis for understanding lung disease, on the other hand. With these concepts in mind, we present a brief synopsis of work in our laboratory to emphasize how evolutionary mechanisms and concepts can be utilized for safe and effective treatment of bronchopulmonary dysplasia (BPD), the chronic lung disease of prematurity, and how these principles can be applied to other biologic systems as well.

CELL–CELL COMMUNICATIONS AS A FRAMEWORK FOR HUMAN EVOLUTION

As Dobzhansky wrote: "Seen in the light of evolution, biology is—perhaps—intellectually the most satisfying and inspiring science. But without this light, it becomes a pile of sundry facts, some of them interesting or curious but making no meaningful picture as a whole" (1973). He aptly concluded that "Nothing in biology makes sense except in the light of evolution." We have focused this "light" to shine on or elucidate *cell–cell communications as a framework for understanding the mechanisms underlying lung evolution as an archetype for human evolution in general, and for explaining lung evolution in particular.*

CANALIZATION, DECANALIZATION, AND THE HOLISTIC APPROACH TO THE PRACTICE OF MEDICINE

The emergence and retention of novel cell–cell communications is equivalent to Waddington's concept of "canalization" (1942). Although this concept may have been elusive and difficult to prove up until now, we propose that this concept is in fact consistent with—and may be the mechanistic analog of—the concept of the stabilization of cell–cell signaling. In contrast, we propose that "decanalization" (Gibson 2009) is associated with disease, unmasking the ancestral signaling pathways as fail-safe mechanisms for the perpetuation of the species. In this chapter we explain how this knowledge of "stabilization and destabilization of cell–cell signaling pathways" should be the basis for the practice of modern clinical medicine. Since this approach is based on, and emphasizes, the stabilization of cell–cell communications, locally (local site involved), regionally (organ), and globally (whole body), this holistic approach is equivalent to the Eastern medicine that addresses the integrated whole and how it can be brought back to its equipoise through biofeedback, herbal medications, acupuncture, and other means.

EXPLOITING LUNG EVOLUTION TO PREVENT AND TREAT CHRONIC LUNG DISEASE

We have reduced lung evolution to the emergence of the lipofibroblast phenotype, and its role in facilitating the stretch-regulated production of lung surfactant as a cellular–molecular selection pressure mechanism. To recapitulate, alveolar epithelial–mesenchymal communication via parathyroid hormone–related protein (PTHrP) and leptin evolved to counteract the evolutionary pressures of gravity, oxygen, infection, and changes in diet. This culminated in the induction of the lipofibroblast, which provided direct protection of the mesoderm against oxidant injury, and indirectly led to an increase in the gas exchange surface area and protection against alveolar atelectasis by augmenting surfactant phospholipid and protein synthesis via the coopertivity of lipofibroblasts with alveolar epithelial cells. During this process, initially surfactant acted to lubricate the swim bladder of fish, followed by the expression of surfactant proteins as antimicrobial peptides in the frog lung, and then as an antiatelectatic factor in the lungs of reptiles and mammals, determined by endodermal and mesodermal interactions. This evolutionary model predicts that injury to either the epithelial cell or the lipofibroblast disrupts the epithelial–mesenchymal crosstalk, which will downregulate the molecular signaling pathways involved in maintaining alveolar homeostasis and lipofibroblast differentation, with the resultant reversion of the lipofibroblast back to its myofibroblast origin. Unlike lipofibroblasts, myofibroblasts cannot promote alveolar epithelial cell growth and differentiation, leading to failed alveolarization, a hallmark of chronic lung diseases such as BPD.

In a series of experiments, we have shown that interference with the cell–cell signaling mechanism causes decreased surfactant production due to loss of the paracrine regulation mechanism, reverting the lung back to its primordial state. Using cell–cell communications to increase lung surfactant and the gas exchange surface area as adaptive strategies against environmental challenges such as increased salinity, bacterial infections, and oxygen during mammalian evolution, we have shown that by modulating the same cell–cell communication mechanisms, we can prevent and/or reverse mammalian lung damage caused by over-exposure to oxygen, infection, and other environmental challenges that led the mammalian lung to its contemporary state in the first place. Since peroxisome proliferator–activated receptor gamma (PPARγ) is the key nuclear transcription factor that determines the lipofibroblastic phenotype, which supports epithelial cell growth and differentiation, we have experimentally determined whether augmenting alveolar PPARγ expression by stimulating it with exogenous PPARγ agonists blocks the failure of alveolarization caused by volutrauma, hyperoxia, infection, and exposure to environmental challenges such as tobacco smoke (Torday and Rehan 2009; Rehan and Torday 2007; Rehan et al. 2009). In fact, we have extensive experimental evidence to show that it does exactly that, under both *in vitro* and *in vivo* conditions. These extensive objective empiric data clearly support the view that exogenous PPARγ agonists can not only prevent and/or halt but also rescue myofibroblast transdifferentiation, effectively preventing oxygen-, volutrauma-, infection-, and smoke-induced inhibition of alveolarization in the developing lung.

LUNG EVOLUTION EXPLAINS THE MAGIC OF CONTINUOUS POSITIVE AIRWAY PRESSURE

From the work reviewed throughout this book, it is clear that the evolutionary science provides deeper insights into human health and disease, allowing a non-traditional way of thinking that helps in designing equally nontraditional and often counterintuitive preventive and therapeutic strategies for human disease. For example, evolutionary knowledge of lung development explains why just supplementing surfactant in a surfactant-lacking extremely premature infant doesn't prevent it from developing chronic lung disease, even when there is increased oxygenation and ventilation following provision of the deficient substance, namely, pulmonary surfactant. This is because chronic lung disease is not simply the lack of surfactant in the alveoli, but more fundamentally, it is due to the lack of fully established homeostatic cell–cell communications in the alveolar wall, which as a result leads to surfactant insufficiency. Therefore, unless the homeostatic alveolar cell–cell communications can be established, regardless of what other treatment is provided, it will not prevent or reverse chronic lung disease. This principle is likely the basis for the success of continuous positive airway pressure (CPAP), which now is a standard approach for managing premature infants with respiratory distress; specifically, provision of just the right amount of alveolar distension stimulates the alveolar epithelial–mesenchymal cell–cell crosstalk via PTHrP and leptin, leading to a more physiologic alveolar milieu, and that is why premature infants supported on CPAP are less likely to develop chronic lung disease. It seems that Gregory and colleagues, who in the late 1960s and early 1970s started to provide CPAP in an attempt to increase functional residual capacity and blood oxygen levels of premature infants with respiratory distress due to surfactant deficiency, unknowingly capitalized on billions of years of lung evolutionary phylogeny and development, and ended up stimulating alveolar epithelial–mesenchymal cell–cell crosstalk via PTHrP and leptin (Gregory et al. 1971).

THE PARADOX OF INFECTING THE LUNG IN ORDER TO TREAT LUNG DISEASE CAUSED BY INFECTION

The lung's response to infection provides another example of the phenomenon described above. Since lung inflammation is a key factor that predisposes the preterm lung to BPD, we have examined whether lipopolysaccharide (LPS), an endotoxin produced by Gram-negative bacteria, affects alveolar cell–cell communications in the process of developing BPD following exposure to infection. Experimentally, on treating fetal rat lung explants in culture with LPS, we found that acutely, which in fact is equivalent to eons on an evolutionary timescale, the expression of key cell–cell communicating proteins such as PTHrP and leptin, and their downstream targets such as the PTHrP receptor, PPARγ, ADRP, and surfactant protein B (SP-B), increased in a dose- and time-dependent manner (Rehan et al. 2007). As reviewed extensively in Chapters 1, 5, and 7, since exposure to infection was one of the main external environmental stresses that drove mammalian lung development, as expected, exposure to LPS resulted in an acute increase in alveolar epithelial–mesenchymal paracrine signaling, mimicking mammalian lung evolution, both ontogenetically and phylogenetically. However, over time the LPS challenge was overwhelmingly beyond the system's evolved capability to deal with the stress of infection, which leads to failed cell–cell communication, ultimately reverting back to its primordial state, characterized by uncoordinated and primitive epithelial–mesenchymal crosstalk, decrease in and/ or absence of lipofibroblasts, increase in myofibroblasts, decrease in alveolar gas exchange surface area, and ventilation perfusion mismatch, all features characteristic of a primitive lung, that is, ending up where it started, a process that has been termed *evolution in reverse*. On the basis of this precept, it is possible that if we expose the developing or developed lung to low-grade inflammation, or expose it to agents that facilitate cell–cell crosstalk, we might stimulate evolutionarily conserved cell–cell-mediated adaptive responses, thereby rendering the lung more injury-tolerant. That is exactly what has been observed, both experimentally and clinically, by Bry et al. (1997) and Watterberg et al. (1996), respectively, when they showed that on exposure to infection, the lung is more mature and more injury-resistant. We and others have extended these observations by showing the paradoxical acute stimulation and chronic suppression of lung maturation on exposure to inflammation, and how continued exposure to low-grade inflammation and/or modulation of PTHrP–leptin signaling pathways could prevent infection-induced lung damage (Rehan et al. 2007). Clearly, without the mechanistic knowledge of how the mammalian lung evolved, it would have made no logical sense to expose the lung to more infection to prevent damage caused by infection. This is, in fact, similar to the logic of active immunity through immunization by injecting microorganism(s) against which protection is sought. However, in the case of the lung, we have carefully and systematically delineated and implicated the ontogenetic and phylogenetic molecular pathways that drove mammalian lung development and homeostasis to its present stage, allowing us to *predict* the cellular responses and signaling pathways involved in this protective response.

EXPLOITING LUNG EVOLUTION TO PREVENT AND TREAT SMOKING-RELATED LUNG DAMAGE

The same principles also explain the well-documented short- and long-term effects of tobacco smoke exposure on lung physiology and pathophysiology, leading to lifelong consequences such as an increased predisposition to asthma in offspring exposed to smoking during pregnancy. It has been unequivocally shown, experimentally, that *in utero* exposure to nicotine, one of the main constituents of tobacco smoke, disrupts specific molecular paracrine communications between the alveolar epithelium and interstitium that are driven by PTHrP and PPARγ, resulting in the reversal of alveolar myofibroblast–lipofibroblast differentiation, and that by molecularly targeting the alveolar epithelial–mesenchymal paracrine communications, nicotine-induced lung injury can be completely averted under both *in vitro* and *in vivo* conditions.

In the examples described above, we have demonstrated that understanding the evolutionary mechanisms that evolved to mediate alveolar mesenchymal–epithelial interactions and the plasticity of mesenchymal cells during mammalian lung evolution provides a functional genomic approach for lung homeostasis and injury repair in the event of adverse stresses. Since modulation of cell–cell communications can effectively lead to the recapitulation of natural signaling mechanisms that have evolved to determine normal lung structure, function, and homeostasis, by utilizing this knowledge, we are able to understand lung pathophysiology better, and design effective preventive and therapeutic strategies.

THE TROJAN HORSE EFFECT OF CANALIZATION

On basis of the objective evidence reviewed above in the context of preventing and treating chronic lung disease, we argue that by integrating evolutionary concepts into day-to-day thinking about other similar clinical problems as adaptive or maladaptive responses, we would be able to ask more innovative and fundamental questions, and make better informed and more rational decisions with respect to the evolutionary origins of organs and disease processes. For example, the concept of canalization as the stabilization of cell–cell communications offers the opportunity to introduce the concept of a Trojan horse effect (see related discussion in Chapter 7, section on vertical integration of leptin signaling, human evolution, and the Trojan horse effect); that is, positive selection of deep homologies can inadvertently introduce either adaptive polymorphisms that explain plasticity and evolvability, or maladaptive polymorphisms that cause disease. However, if these maladaptive polymorphisms are recessive, they will become problematic only if both parents have that same genetic trait (or will not if there are counterbalancing polymorphisms for the trait). Moreover, if the Trojan horse effect introduces a maladaptive phenotype, then in theory one could drive and rectify the process in an adaptive direction by overexpressing the counterbalancing polymorphisms.

Some specific examples of the Trojan horse effect include the increase in vitamin D metabolism and selection for specific P450 isoforms in response to the inhibition of antimicrobial peptides by increased ocean salinity; appearance of $\alpha 3$ type IV collagen in the lung and kidney during the vertebrate transition from water to land; and selection pressure for neocortical evolution, namely, myelinization, which has now resulted in increased lipid deposition in blood vessels, causing coronary heart disease. Inhibition of antimicrobial peptides by increased ocean salinity resulted in inflammation, which led to positive selection pressure via increased host defense by increasing leptin and adaptive changes in vitamin D metabolism and receptor expression; therefore, it is not surprising that now there is a large body of evidence linking vitamin D deficiency to autoimmune diseases. Because of the effect of vitamin D on blood pressure regulation, the increase in vitamin D might have also resulted in a genetic shift in the P450 cytochrome CYP3A from a salt-sensitive form to a salt-resistant form. Vitamin D's Trojan horse effect on blood pressure is supported experimentally by deleting both the vitamin D receptor and 1α-hydroxylase, which activates vitamin D, in mice, resulting in elevated blood pressure, an effect that was reversed by including vitamin D in the animals' diet. In other such studies, the immune response is similarly affected. Since the P450 enzyme system evolved in response to exposure to novel chemicals via plant life, P450 polymorphisms have been linked to the body's ability to metabolize exogenously administered drugs and chemicals, and have also been implicated in the body's response to oxygen and certain P450 enzyme-related diseases.

Another example is the selection pressure for type IV collagen (α3), which acts to physically stent the walls of the lung airsacs, or alveoli. Type IV collagen (α3) isoform prevents the exudation of water and proteins from the microcirculation into the alveolar space. Goodpasture's syndrome, characterized by simultaneous kidney and lung failure, is caused by pathogenic circulating autoantibodies targeted to a set of discontinuous epitope sequences within noncollagenous domain 1 of α3 type IV collagen [α3(IV)NC1]. The α3(IV) chain is not present in worms (*Caenorhabditis elegans*) or flies (*Drosophila melanogaster*), and was first detected in fish (*Danio rerio*). Interestingly, native *Danio rerio* α3(IV)NC1 does not bind to Goodpasture autoantibodies, whereas the recombinant human α3(IV)NC1 domain does. Three-dimensional molecular modeling of the human NC1 domain suggests that evolutionary alteration of electrostatic charge and polarity due to the emergence of critical serine, aspartic acid, and lysine amino acid residues, accompanied by the loss of asparagine and glutamine, contributes to the emergence of the two major Goodpasture epitopes on the human α3(IV)NC1 domain, as it evolved from *D. rerio* over 450 million years. The evolved α3(IV)NC1 domain forms a natural physicochemical barrier against the exudation of serum and proteins from the circulation into the alveoli or glomeruli, due to its hydrophobic and electrostatic properties, respectively, which were more than likely the molecular selection pressure for the evolution of this protein, given the oncotic and physical pressures on the evolving barriers of both the lung and kidney. These evolutionary insights are clearly critical in understanding Goodpasture's syndrome and designing effective therapeutic strategies against this condition, which can be life-threatening.

It is possible that the central role played by fat cells during the development and selection pressure for integrated control of respiration, body temperature, locomotion, and neocortical evolution provides another example of the Trojan horse effect. Fat cells produce leptin, which increases surfactant synthesis, regulates endothermy, and stimulates limb growth. Lipids are also necessary for myelination and neuronal signal transduction, facilitating integrated control of respiration, body temperature, locomotion, brain development, and other evolving physiologic functions. It can be speculated that selection pressure for all of these physiologic functions, including neocortical evolution, if allowed to proceed too far, will result in increased lipid deposition in other organs such as the blood vessels and heart, manifesting as coronary heart disease in type A personality individuals.

IMPETUS FOR EVOLUTIONARY SCIENCE AS AN INTEGRAL PART OF THE CLINICAL CURRICULUM

We advocate that the management of clinical problems, based on evolutionary insights, should be an integral part of any clinical curriculum, rather than simply the trickle-down effects of basic scientists discovering homologies among various phyla. Approaching clinical problems with an evolutionary backdrop lends a perspective that is radically different from—and often counterintuitive to—that based on traditional medicine. For instance, in the examples cited above, the lung responded by overexpressing evolutionarily evolved and conserved mechanisms when challenged with adverse stimuli such as hyperoxia, infection, and tobacco smoke exposure. Similarly, knowledge of how and why bilirubin metabolism evolved during mammalian evolution could potentially explain why it is not a good idea to use phototherapy or exchange transfusion to eliminate all the bilirubin from the neonate's system, even though bilirubin is known to cause bilirubin encephalopathy, kernicterus, liver damage, and bronze baby syndrome. Knowledge of the evolution of bilirubin metabolism not only explains why humans are not green like the many reptilian and amphibian species from which humans evolved, but also the fact that bilirubin is one of the most efficient naturally occuring free-radical scavengers, which on an equimolar basis is probably the most potent antioxidant system that humans have acquired during their transition from water to land in response to high environmental oxygen exposure. Therefore, it should not come as a surprise that for a comparable degree of sickness, infants with relatively higher bilirubin levels have better pulmonary and neurodevelopmental outcomes.

Yet, another example of an evolutionarily conserved response is that of the nutritionally starved fetus that adapts by programming the body into a state that protects against starvation (the thrifty gene hypothesis), and then seems to do "just fine" if continued in the "expected or anticipated" postnatal nutritionally restricted environment. However, if challenged with an "unexpected and unanticipated" nutritionally adequate or surplus environment, the offspring ends up with the deadly syndrome of obesity, diabetes, hypertension, and increased lung disease. We have similarly observed such consequences of maternal food restriction during fetal lung development, which initially results in fewer alveoli in association with decreased alveolar epithelial–mesenchymal (i.e., PTHrP-PPARγ signaling) interactions at birth. However, providing an unrestricted food supply postnatally leads to the initial overexpression of lipofibroblasts, but ultimately to failed alveolar cell–cell communications, surfactant deficiency, and chronic lung disease. Viewed from a descriptive perspective, these effects are difficult to rationalize. But considered as an evolutionary adaptation, this may be a cell physiologic manifestation of the evolutionary energy cost shift toward reproduction that was mentioned back in Chapter 1, resulting in premature breakdown in cell–cell signaling later in life, as reflected by the metabolic syndrome.

APPLICATION OF EVOLUTIONARY SCIENCE TO BIOETHICS

The knowledge of evolutionary science can help in one of the most difficult challenges that clinicians face in clinical practice, namely, bioethics of counseling parents or prospective parents of offspring with major life-threatening defects such as chromosomal anomalies. Often there is disbelief, sadness, and shock over "Why me or why us?" Explanation that these situations are evolutionary strategies to avoid investing in a weak gene pool might help parents make sense out of these perceivably senseless situations. After all, we know that more than 99.9% of oocytes are eliminated to begin with, an evolutionary strategy to weed out the weak zygotes with potential for genetic defects. These explanations can turn unfortunate circumstances that the affected parents face to an opportunity to make positive contributions to society by magnanimously accepting the eventual outcome without any feelings of guilt, based on the laws of mother nature, rather than on the opinion of counselors.

Similarly, we anticipate that in the not very distant future we might be able to make decisions on situations such as brain death, withdrawal of support, and need for organ transplantation on the basis of evolutionary approaches, by, for example, documenting complete lack of cell–cell communications in neurons in the case of brain death, suggesting complete reversal back to the unicellular or close to the unicellular state along the course of vertebrate evolution. Our proposed periodic table of biology might help us in objectively deciding the cutoff for the cellular integrity and the structural and functional reversibility or irreversibility of the clinical situation.

EVOLUTIONARY SCIENCE, A BIOLOGIC PERIODIC TABLE, AND A UNIFIED THEORY OF BIOLOGY

In understanding evolutionary concepts, our broader goal is to decipher the first principles of biology, which will allow us to impart order and hierarchical organization to biological processes, leading to the construction of a biologic periodic table similar to the Periodic Table of Elements, ultimately providing a unified theory of biology. We envision that it could be based on specific functional principles of homeostasis, linked mechanistically through the genes that determine the processes of homeostasis in a wide range of tissues, representing evolutionarily evolved and conserved genes, ligand–receptor pathways, leading to key biologic processes, deciphered through functional and comparative genomics. Then, using interactive algorithms such as self-organizing maps, neural networks, Haplotter, or Protein Analysis Through Evolutionary Relationships (PANTHER) Gene Ontology, we can integrate the genes, ligand–receptor pathways, and the resultant biologic processes, potentially leading to the generation of a biologic periodic table. Citing the example of the cell–cell interactions involved in the evolution of the lung alveolus, in Chapter 8 we have proposed a framework for mathematical modeling that integrates the genes for the ligand–receptor pathways involved in surfactant synthesis, and the resultant alveolar development involving the progression from associative, to additive, to multiplicative, to exponential, and finally to combinatorial/permutative relationships of genes involved in alveolar epithelial–

mesodermal crosstalk. Once similar relationships are discovered and proved to hold true in other organ systems, such as the eye, heart, liver, and kidney, it would provide a novel paradigm for practicing *predictive, preventive, and therapeutic medicine,* based on genes and signaling pathways evolved for homeostasis, injury repair, and survival under contemporary environments and lifestyles. Our argument, elaborated in Chapter 8, is that functional and structural homology to cholesterol could well serve as a central organizing principle for assigning position and valence to other elements of the biologic periodic table. This argument is further supported by the work from Voight and colleagues (2006), who have shown that some of the genes involved in most recent human evolution included genes involved in the processing of dietary fatty acids, including uptake (*SLC27A4* and *PPARD*), oxidation (*SLC25A20*), and regulation (NCOA1 and *LEPR*). The latter gene (*LEPR*) is the leptin receptor, which plays an important role in regulating adipose tissue mass. Furthermore, genes in the phosphatidylinositol pathway, which mediates leptin receptor signaling, were also particularly overrepresented among the significant genes involved in most recent human evolution, including *INPP5E, PI4K2B, IHPK1, IHPK2, IHPK3, IMPA2,* and *SYNJ1.*

SUMMARY

In summary, we suggest an evolutionary approach, based on cell–cell communications and functional cellular and molecular compartmentalization, to all problems in clinical medicine. We rationalize that cell–cell communication-specific mechanisms that have resulted in evolution of human lineages under selection pressure can be exploited to understand both homeostasis, representing health, and the breakdown in homeostasis, representing disease. These principles should form the basis for modern clinical science, and we urge clinicians to step up to the plate now, rather than wait for the creation of billing codes such as "chronic lung disease caused by failed cell–cell communication and maladaptive host response resulting in reversal to primordial lung" or "failed postnatal adaptive response leading to the syndrome of obesity, diabetes, hypertension, and increased lung disease" to appear before being forced to incorporate evolutionary medicine into clinical practice. Bear in mind that it took 1500 years to transition from Ptolemy's Earth-centric to Copernicus' heliocentric model of the solar system.

REFERENCES

Bry K, Lappalainen U, Hallman M (1997), Intraamniotic interleukin-1 accelerates surfactant protein synthesis in fetal rabbits and improves lung stability after premature birth. *J. Clin. Invest.* 99:2992–2999.

Dobzhansky T (1973), Nothing in biology makes sense except in the light of evolution. *Am. Biol. Teacher* 35:125–129.

Gibson G (2009), Decanalization and the origin of complex disease. *Nature Rev. Genet.* 10(2):134–140.

Gregory GA, Kitterman JA, Phibbs RH, Tooley WH, Hamilton WK (1971), Treatment of the idiopathic respiratory distress syndrome with continuous positive airway pressure. *N. Engl. J. Med.* 284:1333–1340.

Rehan VK, Torday JS (2007), Exploiting the PTHrP signaling pathway to treat chronic lung disease. *Drugs Today (Barcelona).* 43(5):317–331.

Rehan VK, Dargan-Batra SK, Wang Y, Cerny L, Sakurai R, Santos J, Beloosesky R, Gayle D, Torday JS (2007), A paradoxical temporal response of the PTHrP/PPARgamma signaling pathway to lipopolysaccharide in an in vitro model of the developing rat lung. *Am. J. Physiol. Lung Cell. Mol. Physiol.* 293(1):L182–L190.

Rehan VK, Asotra K, Torday JS (2009), The effects of smoking on the developing lung: Insights from a biologic model for lung development, homeostasis, and repair. *Lung* 187(5):281–289.

Torday JS, Rehan VK (2009), Exploiting cellular-developmental evolution as the scientific basis for preventive medicine. *Med. Hypotheses* 72(5):596–602.

Voight BF, Kudaravalli S, Wen X, Pritchard JK (2006), A map of recent positive selection in the human genome. *PLoS Biol.* 4(3):446–458.

Waddington CH (1942), Canalization of development and the inheritance of acquired characters. *Nature* 150:563–565.

Watterberg KL, Demers LM, Scott SM, Murphy S (1996), Chorioamnionitis and early lung inflammation in infants in whom bronchopulmonary dysplasia develops. *Pediatrics* 97:210–215.

NAME INDEX

Adamson, I. 46
Arthur, W. 12, 56
Avery, M. 18, 46, 78, 92

Bernard, C. 67
Berner, R. 83, 107
Bloch, K. 26, 78, 101
Bloom, M. 7, 102
Bonner, J. 49

Carter, B. 124
Cavalier-Smith, T. 7, 101, 102
Clements, J. 78, 100, 128
Cope, E. 66
Crespi, E. 43, 84
Csete, M., 83, 84, 107

Daniels, C. 31, 41, 100
Darwin, C. 12, 25, 26, 36–38, 55, 60,
   67, 76, 78, 95, 96, 112, 146
Denver, R. 43, 84
Dobzhansky, T. 43, 56, 140

Einstein, A. 37, 38, 40, 56, 76, 116

Faridy, E. 18
Fine, L. 44, 126

Gehring, W. 78
Gerhart, J. 67
Gould, S. 81
Grobstein, C. 17

Horowitz, N. 39, 78, 100, 101, 121

Jablonka, E. 36
Jantsch, E. 66
Jaskoll, T. 49

King, J. 46
King, N. 12
Kipling, R. 126
Kirkwood, T. 14
Kirschner, M. 67
Kosswig, C. 56
Kotas, R. 46
Krogh, A. 29, 131

Lamarck, J.-B. 54, 56
Lamb, M. 36
Lewis, E. 65
Liggins, G. 18
Linnaeus, C. 36, 43, 79, 115

Mandelbrot, B. 67
Margulis, L. 4, 10
Mayr, E. 56
Medawar, P. 13
Melnick, M. 49
Mendeleev, D. 76, 115, 116,
   120, 121
Michaelson, A. 38
Morley, E. 38
Morowitz, H. 3, 4
Mouritsen, O. 7, 102
Mullis, K. 67

Orgeig, S. 31, 41, 100

Evolutionary Biology, Cell–Cell Communication, and Complex Disease, First Edition.
John S. Torday and Virender K. Rehan.
© 2012 Wiley-Blackwell. Published 2012 by John Wiley & Sons, Inc.

SUBJECT INDEX

Evolutionary Biology, Cell–Cell Communication, and Complex Disease, First Edition.
John S. Torday and Virender K. Rehan.
© 2012 Wiley-Blackwell. Published 2012 by John Wiley & Sons, Inc.